MINING THE NATION'S BRAIN TRUST

MINING THE NATION'S BRAIN TRUST

HOW TO PUT FEDERALLY-FUNDED RESEARCH TO WORK FOR YOU

FRED E. GRISSOM, JR. AND RICHARD L. CHAPMAN

ADDISON-WESLEY PUBLISHING COMPANY, INC.
Reading, Massachusetts   Menlo Park, California   New York   Don Mills, Ontario
Wokingham, England   Amsterdam   Bonn   Paris   Milan   Madrid   Sydney   Singapore
Tokyo   Seoul   Taipei   Mexico City   San Juan

Many of the designations used by manufacturers and sellers to distinguish their products are claimed as trademarks. Where those designations appear in this book and Addison-Wesley was aware of a trademark claim, the designations have been printed in initial capital letters.

The publisher offers discounts on this book when ordered in quantity for special sales. For more information please contact:
Corporate & Professional Publishing Group
Addison-Wesley Publishing Company
One Jacob Way
Reading, Massachusetts 01867

Grissom, Fred E.
  Mining the nation's brain trust: how to put federally-funded research to work for you / Fred E. Grissom, Jr. and Richard L. Chapman.
      p.   cm.
    ISBN 0-201-55015-6
    1. Technology transfer—United States. 2. Federal aid to research—United States. I. Chapman, Richard L. II. Title.
T174.3.G77   1992
338.97306—dc20   92-14167
                                                              CIP

Copyright © 1992 by Addison-Wesley Publishing Company, Inc.

Cover design by Simone R. Payment

   All rights reserved. No part of this publication may be reproduced, stored in a retrieval system, or transmitted, in any form or by any means, electronic, mechanical, photocopying, recording, or otherwise, without the prior written permission of the publisher. Printed in the United States of America. Published simultaneously in Canada.

ISBN 0-201-55015-6
1 2 3 4 5 6 7 8 9 10   MU   9695949392
First Printing, July 1992

To the scientists and engineers, both civil service and contractors, who daily press forward with dedication and an innovative spirit in the service of their country. Any American who visits one of the hundreds of federal laboratories cannot help but be thrilled and proud of the exciting developments being produced there, with their powerful potential to further human achievement and strengthen the U.S. economy.
 —From the authors

To my parents, Fred and Juanita, and sisters, Jane, Sally, and Rebecca
 —Fred Grissom

TABLE OF CONTENTS

	Preface ix
Part I	**The Bottom Line: Why Take Federal Technology Transfer Seriously?** 1
Chapter 1	Money Is Being Made and Saved 3
Chapter 2	Federal Technology: What's in It for You? 7
Chapter 3	So, What Has Changed? Technology Transfer Has Become a Priority for Federal Laboratories 13
Chapter 4	Strong Incentives for Cooperation 19
Chapter 5	How Are the Agencies Organizing to Respond? Agency Perspectives and Management Support 25
Chapter 6	The Laboratories 37
Chapter 7	Yes, But What About...? The Myths and Realities of Technology Transfer 43
Chapter 8	Successful Technology Transfer: Lessons Learned 49
Part II	**Where Can You Look for Help?** 53
Chapter 9	Brokers 55
Chapter 10	Technology Transfer by Information Dissemination 63
Part III	**Obtaining Existing Technology** 73
Chapter 11	Patent Licensing 75
Chapter 12	Software 83
Chapter 13	Contractor-Owned Technology 87
Chapter 14	Spinoff Companies and Employees 91
Part IV	**Working Together to Create New Technology** 95
Chapter 15	Cooperative Research and Development Agreements 97
Chapter 16	Use of Unique Laboratory Facilities 101
Chapter 17	Other Mechanisms for Working Together 105
Part V	**Epilogue** 113
Appendices	
I.	Technology Transfer Contacts 119
II.	Technology Transfer Checklist 149
III.	Sample Documents 157

PREFACE

This book outlines the opportunities and pitfalls associated with the transfer of technology from federally funded research and development to the private sector. In simple, straightforward language, using numerous examples, we show how federal technology can be transferred to commercial use—with considerable benefit to the cooperating parties.

Progress over the past decade offers many lessons to both government and industry on how this process can be successful. Congress has been active in promoting technology transfer through a variety of enabling legislation. Implementation of this legislation has accelerated, with increasing awareness and innovation in federal laboratories.

As you will discover, the payoff for successful technology transfer can be high to federal scientists, to industrial cooperators, and to private entrepreneurs. But there needs to be a better understanding among all parties about the rules of the game. This process calls for a new and different relationship between government and the private sector. It requires a higher degree of trust and cooperation than may have existed in the past. In many respects this process remains experimental and involves both institutional and personal risk.

Our purpose is to provide practical guidance, based on experience, for government, industry, and other technology managers about how to take full advantage of mining the nation's brain trust.

We acknowledge the support and assistance of our wives, Marilyn Chapman and Shelley Grissom; the secretarial assistance of Jody Briles; the editorial assistance of Bonnie Kate Brown; for believing in the project from the beginning, Ted Buswick; for shepherding this work through the intricacies of the publishing process, the staff at Addison-Wesley, especially Leslie Morgan and Jennifer Joss; and aerospace colleagues who are committed to technology transfer and have generously shared their experiences.

Richard Chapman and Fred Grissom

PART I

THE BOTTOM LINE
WHY TAKE FEDERAL TECHNOLOGY TRANSFER SERIOUSLY?

CHAPTER 1

MONEY IS BEING MADE AND SAVED

There are innumerable examples of new and improved products or savings in process and products stemming from federal laboratory research, both indirectly and in specific collaborations with industry to produce new products. This collaborative process has been greatly stimulated by the Technology Transfer Act of 1986, which opened up a wide variety of avenues for cooperation between government laboratory scientists and industry. A few examples will help you understand the potential that rests here—both in terms of magnitude and variety.

NEW AND IMPROVED PRODUCTS

The National Renewable Energy Laboratory (NREL), one of the major laboratories of the U.S. Department of Energy (DOE), undertook the development of a new fiberglass blade designed specifically to meet the operating requirements of horizontal-axis wind turbines that produce electrical energy. Initial tests showed that these new blades could slash wind energy costs by as much as 20 percent. In cooperation with several private firms, NREL is conducting further on-site tests. Turbine blades for the more than thirty-six hundred wind turbines in California will have to be replaced within the next decade, providing a potential market in excess of $30 million for U.S. manufacturers.

In an especially dramatic example, the Department of Veterans Affairs (VA), working in cooperation with the National Aeronautics and Space Administration (NASA), Washington University in St. Louis, the Central Institute for the Deaf, and the Minnesota Mining and Manufacturing Company, is developing the adaptive digital hearing aid. This new development promises to revolutionize the hearing aid market as it provides the flexibility of adjustment and control of auditory signal processing required to match an individual person's hearing impairment.

Going back some forty years, the Agricultural Research Service (ARS) was faced with the growing problem that potato consumption was declining in the United States. An ARS research team came up with the necessary steps to process potatoes into instant potato flakes—helping to boost a new element of the potato industry to what is now a $400 million a year market.

More recently ARS scientists have developed a formulation from corn starch which can absorb nearly fourteen hundred times its own weight in water. Dubbed "super slurper," it is used today in body powders, absorption dressings, soil additives, diapers, batteries, and oil/petroleum filters.

One of NASA's primary research centers, experimenting with coating technology for a possible use in astronaut helmet visors, developed, tested, and licensed a particularly long-lasting coating. Today a major manufacturer of eye wear is using a derivative of this technology to make scratch-resistant lenses, producing a market in excess of $50 million per year.

There has been a virtual explosion in cooperation with industry emanating from the National Institutes of Health (NIH), which in 1990 filed patents on 300 different inventions. In many instances further work will be done in cooperation with drug and biological firms for use in testing, disease diagnostics, and disease treatment.

Medical science has also progressed rapidly in medical diagnosis based on the technology of digital image processing, much of the basic work having been accomplished in NASA laboratories. Again, the applications of this technology, for instance in CAT-scan and MRI diagnostics, are multiplying and its further development is exploding through government-industry cooperation.

SAVINGS IN PROCESS AND PRODUCTS

Recently, the Air Force Engineering and Service Laboratory patented a new method for treating industrial waste water using ferrous sulfate/sodium sulfide treatment. This new means of treating waste water metals reduces quantities of hazardous sludge, eliminates the need to soften water (for reuse), and is estimated to produce savings of about $655,000 per year in a typical 1.4 million gallon per day waste treatment system.

During the Apollo days, NASA laboratories produced a structural analysis software program, termed NASTRAN, which has been widely used and further developed in everything from automobile design to ship construction. Literally hundreds of variations can be found throughout industry today. A single application has resulted in design cost savings amounting to millions of dollars.

The Environmental Protection Agency (EPA), working in cooperation with industry, has developed a program to test and verify the various means for toxic waste remediation techniques. This is crucial for standard setting in an arena where potential litigation can be extremely costly and where there has been a dearth of standards. It promises to save industry, to say nothing of the U.S. public, hundreds of millions of dollars in future years.

The DOE recently announced the establishment of an industry consortium consisting of wood products, chemical, plastics, and high technology companies to exploit a DOE-developed technology for converting wood waste to phenol and phenolic resins. The market for such resins totals about 2.7 billion pounds per year. The newly developed process using biomass as a principal feed stock (in contrast to petroleum) is estimated to provide adhesives in the price range of sixteen to twenty-seven cents per pound compared to petroleum-derived adhesives at forty-five cents per pound.

The Army Corps of Engineers Civil Engineering Research Laboratories (CERL) conducts research to solve problems and provide solutions to military engineers. Frequently, they develop an item that must be transferred to industry so that it can be produced in quantity for the military—often resulting in widespread civil use as well. This was the case when CERL developed the ceramic anode, a high technology anode for cathodic protection of civil work structures such as underground pipes, water towers, and the like. This new device sharply reduces corrosion, can be produced and installed at half the cost, and is only 1/500 the weight of the more commonly used silicon-iron anodes.

These few illustrations, ranging from the rather common to the more exotic, give insight into the potential payoff from mining the nations's brain trust in federally sponsored technology.

CHAPTER 2

FEDERAL TECHNOLOGY: WHAT'S IN IT FOR YOU?

This question can be answered best in three stages: first, there is the immense collection of resources poured into producing federal technology; second, there is evidence that this technology is useful and can be located and acquired; third, real benefit can accrue to those who participate in this technology transfer effort.

THE MAGNITUDE OF FEDERAL RESEARCH AND DEVELOPMENT

The federal government is the largest supporter of technological research and development (R & D) in the United States, and much of the technology developed by federal agencies has potential commercial use. In fiscal year 1990 nearly $70 billion was allocated to R & D by all federal agencies, including both contractor and in-house activity. Only 15 percent of the federal R & D budget is spent in-house; the remaining 85 percent goes to both private contractors and nonprofit research centers. In total it represents nearly one-half of all funds spent in the United States on R & D.

Federal agencies have 400 laboratories across the country, each staffed by ten or more scientists and engineers. More than fifty laboratories have staffs of one thousand or more. Some 200,000 staff provide the talent for these labs. This is a tremendous effort.

More than 700 patents are awarded and over 900 patent applications are made annually by federal agencies. In 1989, 34 of the 100 R & D winners were from the federal labs. In 1990 the number increased to 39, and this is just the tip of the iceberg. A review of the new technology disclosures published in *NASA Tech Briefs* magazine revealed that up to 80 percent of new technology reported by the federal government is not patented. On that basis one can estimate that there are approximately ten thousand new discoveries annually that occur through federally supported R & D.

In summary, there is a wealth of information, talent, and unique facilities supported by federal research and development dollars—the vast bulk of which can become available to U.S. industry on a mutually beneficial basis—*if* one is willing to apply some systematic effort.

For example, virtually every agency that supports or conducts research has a significant data base holding information about research completed and, often, research in progress. Most federal agencies report this information to the National Technical Information Service (NTIS), a governmentwide repository operated by the Department of Commerce. Currently, the NTIS claims to have more than 2 million entries of federally conducted or sponsored research.

The one hundred thousand scientists and engineers employed by the federal government in its research facilities nationwide constitute an extraordinary source of talent. Their expertise ranges across all possible topics and interests to include such endeavors as experimental surgery, flight dynamics, magnetic resonance imaging, physics of materials, molecular biology, and toxicology. The federal government conducts or supports work on virtually every topic of legitimate scientific or engineering interest.

Finally, with its wide array of laboratories comes a similar array of facilities that often are unique. Typically, these facilities can be made available to outside investigators on a noninterference basis. These include such facilities as hyperbaric chambers, a wide variety of wind tunnels, the national synchrotron light source, the national germ plasm collection, and large-scale hydrology models.

This collection of data bases, talent, and facilities constitutes a tremendous series of assets, potentially available to industry researchers.

THIS LODE CAN BE EFFECTIVELY MINED

Virtually from the beginning of the republic, the U.S. government has supported research or exploration activities for both specific and broad public purposes. But usually it has left the application to private initiative. For example, in 1803 Congress appropriated $2,500 for the Lewis and Clark expedition, which opened up the Great Northwest. The reports and maps resulting from this expedition provided information for an army of entrepreneurially minded individuals who later exploited this effort.

In 1831 the U.S. government financed a series of experiments by the Franklin Institute in Philadelphia to determine why boilers on steamships were exploding with disastrous results. The institute *did not* build new boiler systems—but they provided the technology.

In 1845 Congress made available $30,000 to build a prototype telegraph system for testing between Washington, D.C., and Baltimore, Maryland. Again, the government did not enter the telegraphy business—it provided the wherewithal for testing that permitted further development toward a commercial reality.

In 1862 the vast land grant university system was established, and along with it began the system of agricultural research that ranged from basic research to developmental activities—all of which provided technology to what we now call the agribusiness community.

Over the past two hundred years federally financed R & D has multiplied greatly, but application does not seem to have kept pace. Knowledgeable observers believe that substantial opportunities await those who are willing to mine the current lode of federally sponsored technology. Recently, the president of SRI International noted that the United States is on the verge of a major wave of technological exploitation. He stated that many of the technologies of the 1960s and early 1970s were based on prewar science. There is a big backlog of postwar science now maturing. We are seeing just the leading edge of its commercialization.

Admittedly in the past, searching for and extracting technology from federally funded efforts have been real chores. However, during the past decade a series of legislative initiatives have facilitated the successful transfer of technology from federal laboratories to other spinoff purposes. Central to this effort are four key pieces of legislation and a presidential executive order.

P.L. 96-480, the Technology Innovation Act of 1980 (often called the Stevenson-Wydler Act), encourages agencies to provide offices and staff for research and technology applications and to help industry and other potential users gain easier access to federally sponsored technology. This legislation *permitted* but *did not require* agencies to fulfill this function.

P.L. 96-517, the Patent and Trademark Amendments of 1980 (often referred to as the Bayh-Dole Act), provides easier access to patent rights resulting from federally sponsored R & D. Basically, this act permits small businesses, universities, and other not-for-profit organizations to have first call on technology developed with federal R & D funds for possible patenting purposes.

Perhaps the key piece of legislation is P.L. 99-502, the Technology Transfer Act of 1986. This act changed the emphasis from *permitting* agencies to open up technology transfer activities to *requiring* agencies to work more closely with industry for the successful transfer of technology.

This act was further strengthened when President Reagan issued Executive Order 12591 (April 10, 1987), "Facilitating Access to Science and Technology." The executive order went a step further by encouraging agencies to seek third parties to assist in this process. It also authorized cooperative research ventures of consortia of laboratories, industrial companies, and universities to which federal agencies could actually make cash contributions. Further, it required the Department of Defense (DOD) to make special efforts to overcome the barrier of security classification.

Finally, P.L. 101-189, the National Competitiveness Technology Transfer Act of 1989 was passed as part of the DOD 1990 Authorization Act. It further encouraged technology transfer by opening up the government-owned, contractor-operated (GOCO) national laboratories. Through these great laboratories, primarily under the aegis of the Department of Energy, industry can tap important, unique, cutting-edge technology.

Can positive technology transfer benefit companies that make the effort? Absolutely!

For example, ITT Gilfillan, a unit of ITT's Defense Technology Corporation, recently reported savings of up to $1 million per year through the use of structural adhesives technology that was made available to the company by an Industrial Applications Center's (IAC) search and screening of many data bases. The IACs, recently renamed Regional Technology Transfer Centers (RTTCs), were part of a NASA-supported network dedicated to improving the technology transfer process.

In a quite different instance, technology transfer resulted in a new product development based on federally supported technology. When Ray Ward moved from Utah to Arizona, he found he did not like the taste of the water. A tenacious innovator, he explored a variety of filtering systems and learned of a unit that had been developed to purify water aboard the space shuttle Orbiter. After receiving a technical information packet from NASA, he developed a filter that satisfied his purposes. Although he hadn't planned to enter the manufacturing business, requests from friends, neighbors, and other people launched a business enterprise, incorporated in 1977 as Bon Del Manufacturing Company, which ultimately became Water Filter Company of America, a subsidiary of Dubarry, and resulted in a multimillion-dollar-a-year business.

The payoffs are large for those who participate!

THERE ARE BENEFITS FOR ALL PARTICIPANTS

For industry, cooperation is essential to successful technology transfer that will broaden the technological base. This is particularly important today. No matter what size, market share, or research investment a company attains, technology development is moving too rapidly and has become too complex for any single company, no longer respecting either corporate or national boundaries.

Technology transfer accelerates technological payoffs—companies can obtain a running start by avoiding R & D dead ends. Technology transfer also permits a company to bypass costly duplication when current problems have been solved elsewhere.

Effective technology transfer can produce dramatic results. For example, a recent study of the spinoff applications of NASA technology revealed that only 259 such applications contributed nearly $22 billion to sales or savings by the users ("An Exploration of Benefits from NASA Spinoffs," Chapman Research Group, Inc., June 1989, Littleton, CO).

Individual scientists and engineers can benefit as well. They grow, develop, and improve their skills in several ways: (1) they can have contacts with peers that may not otherwise have been available; (2) they may extend their knowledge of both their disciplines *and* previously inaccessible applications of research; and finally, (3) they may receive royalty income from licensing such technology. For example, in cooperative efforts between federal laboratories and industry, participating federal scientists and engineers can earn as much as $100,000 per year from royalties on their inventions in addition to their regular salaries.

Federal agencies that participate in technology transfer also significantly benefit. Like the individual scientists or engineers, agencies profit from the broader context and access to knowledge outside their regular orbit, including the greater variety of applications and application environments. Exposure to other kinds of scientific or engineering circumstances has a synergistic effect in stimulating innovative thought. Participating agencies can supplement their income in cooperative ventures by accepting funding from outside sources in such joint efforts. They also may receive additional political support for their research programs from satisfied collaborators.

Finally, the nation benefits because technology transfer permits the more expeditious exploitation of innovation, a new synergism from cooperation between industry and government, an improved status in the increasingly intense international economic and scientific competition, and practical additions to the economy. In the study already noted, the spinoff applications also resulted in estimated additional revenues from the federal corporate income tax of nearly $356 million. But more important, more than 350,000 jobs were created or saved through these technology transfer applications.

In summing up, what's in federal technology for you? There can be a great deal—especially if you are willing to learn how to make the system work *for* you.

CHAPTER 3

SO, WHAT HAS CHANGED? TECHNOLOGY TRANSFER HAS BECOME A PRIORITY FOR FEDERAL LABORATORIES

TECHNOLOGY TRANSFER IN HISTORICAL PERSPECTIVE

The term *technology transfer* is relatively new, usually dated back to the Technology Utilization Program established by NASA in the early 1960s. However, the longest, most continuous, organized effort began in 1862 with the creation of the Department of Agriculture and the establishment of the land grant university system throughout the United States. The purpose of the latter was to provide for colleges of agriculture and mechanical arts that would stimulate both agriculture and commerce. More specifically, the new department was created "to acquire and diffuse...useful information on subjects connected with agriculture...to procure, propagate, and to distribute among the people new and valuable seeds and plants...and to conduct practical and scientific experiments" (the Morrill Act of 1862). Shortly thereafter, the new department began to conduct research in chemistry, botany, and entomology. By the mid 1870s several states had established agricultural experiment stations at the land grant institutions. The Hatch

Act of 1887 authorized federal funding of the state-developed experiment stations. Two years later the Department of Agriculture was given cabinet status. In 1914 the Smith-Lever Act provided for a separate extension service to promote technology transfer, and in 1935 the Bankhead-Jones Act unified agricultural research.

Other examples of continuing research serving such useful purposes as promoting the transfer of technology to civil purposes included the establishment of meteorological work in the Army Signal Corps in 1870; the establishment of the U.S. Geological Survey in 1879; and the development in 1887 of the Hygienic Laboratory as part of the Marine Hospital Service in New York, which eventually grew into the National Institutes of Health (NIH). All of these activities required active mechanisms to transfer research from the laboratory or proving grounds into daily commerce, medical practice, agriculture, navigation, and so on to meet public purposes. Each activity served specific purposes and clientele involving direct or *vertical* technology transfer.

NASA in 1963 established its Technology Utilization Program to promote the commercialization of technology based on its aeronautical and space research and development (R & D). The primary difference between this kind of technology transfer and that experienced from earlier organizations (or even from NASA's predecessor, the National Advisory Commission on Aeronautics) was that this effort was directed toward the secondary use of technology—characterized as *horizontal* transfer because it moved laterally *across* organizational boundaries. *This meant facilitating the use of technology originally developed for one purpose or location in a quite different setting.* For nearly ten years the NASA program was the only specifically organized program to promote this type of spinoff technology transfer.

Then, beginning in 1971, a group of laboratories in the Department of Defense (DOD) established the DOD Technology Transfer Laboratory Consortium, which originally transferred technology on an interagency basis among laboratories within DOD. Later it began to provide technical assistance to state and local governments, and by mid-1976 the Federal Laboratory Consortium (FLC), now consisting of laboratories from the civil agencies as well, established a secretariat function and was reaching out to everyone interested in using federally developed technology. With the subsequent passage of legislation noted in Chapter 2, federal laboratories and their mother agencies have been given both the authority and the mandate to facilitate the domestic, civil use of federally sponsored technology to provide new and improved products and processes, and generally to strengthen U.S. industry.

The primary policy for technology transfer in federal laboratories was laid down in the Technology Transfer Act of 1986 (P.L. 99-502), which mandated that agencies having significant research activities in-house would act vigorously to transfer technology so as to commercialize such technology and, thereby, strengthen the economic competitiveness of U.S. industry. This act substantially extended and strengthened the original technology transfer legislation, the Stevenson-Wydler Technology Innovation Act of 1980, which had acknowledged the value of technology transfer, recognized it as an important function, but tended only to legitimize grass roots technology transfer efforts being conducted at that time. The 1980 act had merely required that each federal agency laboratory with an annual budget of $20 million or more establish an Office of Research and Technology Applications (ORTA). Sadly, most agencies paid little attention to this legislation, and there were no substantial means for enforcing it. On the other hand, the Technology Transfer Act of 1986 provided new tools to federal laboratory managers and stiffened the requirements, clearly making technology transfer more important within the federal R & D community.

Perhaps the most important new authority in the act was section 11, which provided for cooperative research and development agreements (CRADAs). This new authority gave the laboratory director permission to enter CRADAs with virtually any kind of organization, provided that the activity fell within the mission of the laboratory. Departments and agencies were specifically required to delegate this authority to their laboratory directors. Laboratories were authorized to accept, retain, and use funds, personnel, services, and property as a part of these cooperative R & D agreements. In turn, the laboratory could furnish personnel, services, and property (but not funds) for such endeavors.

Recognizing the challenges faced by private enterprise in commercializing technical discoveries, further authority was granted to the laboratories to arrange *in advance* for patent licensing arrangements within these CRADAs by which laboratory directors could waive the government's ownership in patents and licenses *if that would facilitate commercialization of inventions*.

Section 10 of the 1986 act addressed several important changes in the policy facilitating use of federal technology.

First, it made technology transfer a responsibility of each laboratory science and engineering professional, and it expanded this to include a requirement that the technology transfer function be a part of each of these individuals' position descriptions. It was also to be included in performance evaluations for promotion.

Second, it clarified the requirement for the staffing of laboratory ORTAs so that any laboratory having 200 or more full-time equivalent scientists, engineers, or technical personnel would have to devote a full-time position to this function.

Third, the act specified that laboratories were to participate in regional, state, and local programs of technology transfer specifically for the benefit of their region, state, or locality—thus giving laboratories a legitimate role in local economic development.

Fourth, it formally recognized the FLC and provided funds to administer this organization so that it could facilitate and coordinate technology transfer across the federal government.

This push to make technology transfer a stronger priority in federal laboratories was given further emphasis in April 1987 when President Reagan issued Executive Order 12591, solidly placing presidential support behind technology transfer. Basically, the executive order reemphasized those portions of the act calling for collaboration with the private sector. It extended the authority from Section 10 of the act which encouraged laboratories to participate in regional consortia by establishing a new Technology Share Program. The executive order required the secretaries of the Departments of Agriculture, Commerce, Energy, and Health and Human Services, and the administrator of NASA, each to select one or more laboratories to be the focal point for using their particular areas of research expertise to enhance long-term national economic competitiveness through establishment of consortia that would include both U.S. industries and universities. Laboratories could use personnel, facilities, and even contribute funds up to a maximum of $5 million per year to such consortia.

In several areas the executive order went beyond the act. First it required the head of each agency to identify and encourage persons to act as conduits for the transfer of technology developed from federally funded R & D efforts. In other words, it legitimized the role of a technology broker or other third party to facilitate technology transfer between a federal laboratory and a potential user. Further, it required the secretary of defense to take appropriate steps to improve technology transfer to the civil sector.

Just three years after passage of the 1986 act, a further amendment to facilitate technology transfer was made through the National Competitiveness Technology Transfer Act of 1989 (P.L. 101-189), which extended the basic provisions of the 1986 act to permit government-owned, contractor-operated (GOCO) federal laboratories the same authorities that had been given to government-owned and operated (GOGO) laboratories. Probably most important was the authority to enter CRADAs, permitting the director of a GOCO laboratory to enter CRADAs with private sector entities.

The 1989 act expanded the 1986 act, generally, to help protect proprietary information shared in such CRADAs. It specifically authorized that certain technical data could be protected from public dissemination for up to five years under such agreements.

Some agencies have acted rapidly to exploit these authorities. Others have been slower to act, but there is strong evidence that all of the agencies are moving much more rapidly today to find and undertake cooperative efforts with the private sector in commercializing federal technology.

The House Committee on Science, Space, and Technology held hearings in May 1990 to assess progress on the transfer of technology from federal laboratories. Senior officials from the U.S. Air Force, Department of Veterans Affairs (VA), NIH, Environmental Protection Agency (EPA), and the Department of Energy (DOE) all testified to significant and exciting progress under these authorities. Increasingly, technology transfer is being incorporated into the institutional activities of federal laboratories. For example, Air Force Regulation 80-27, Domestic Technology Transfer has formally included technology transfer within the mission of each air force laboratory.

Witnesses testified to a wide variety of problem-solving to assist U.S. industry or develop new and useful products. Agencies are producing special periodicals, flash notices sent to interested industries, handbooks, directories, and are providing further information on technology transfer opportunities through conferences and symposia specifically directed toward industry.

Federal agencies essentially have doubled the number of CRADAs that they have entered or are negotiating each year for the past three years. The NIH alone has more than 200 such agreements.

Federal agencies are getting serious about technology transfer!

CHAPTER 4

Strong Incentives for Cooperation

The recent legislative initiatives to foster technology transfer, especially from federally sponsored technology to commercialization in U.S. industry, have provided strong incentives for cooperation between industry and government. For example, among incentives to industry are the following: (1) a systematic, fair access to federally developed technology; (2) the opportunity for truly cooperative research ventures; (3) the serious potential to reduce R & D costs by turning to sources of government technology; (4) the use of unique facilities, quality researchers, and an extensive store of knowledge; and (5) new licensing opportunities. On the other hand, more attractive incentives for government to cooperate with industry include: (1) a rationale and support of agency R & D programs, (2) new sources (external) of funding for agency R & D, (3) access to otherwise restricted technology, and (4) both institutional and individual rewards in terms of royalty sharing.

Incentives to Industry

For the first time U.S. industry can have access in a systematic and fair manner to federally developed technology. It no longer needs to be random because there is now a general framework and system. Clearly, every agency is somewhat different, but the general framework has similar characteristics and procedures so that the scientist or manager in industry is not required to learn a different track for each agency contacted. As these systems mature, agencies can be expected to introduce even more user-friendly ways to access their technology.

Access is fairer now because success in achieving access and making technology transfers depends less on the right institutional or political connections, and more on the capability of the organization to screen, understand, and use the technology being sought. Basically, there are now common rules across the government that control the process and specifically promote successful transfer.

These two elements are further supported by an increasing investment on the part of government agencies to improve and facilitate the technology transfer process. The recent General Accounting Office (GAO) survey about technology transfer activities in federal agencies, conducted in 1990, revealed expenditures or the use of resources valued at more than $500 million annually devoted to an extensive series of ways to transfer technology. These resources are increasing yearly as more agencies add to their Office of Research and Technology (ORTA) staffs, devise easier means of access to technical data bases, and become more active in advertising their technical capabilities to U.S. industry.

The authority for federal laboratories to enter into cooperative research and development agreements (CRADAs) opens a significant opportunity for industrial access to technology that was quite limited previously. Very few federal agencies had this type of authority (one notable example was NASA). This authority specifically was designed to provide a formal but simple means for agreement on how to conduct a cooperative R & D effort. It is outside the regular procurement function, and CRADAs *are not* procurement instruments. This type of cooperative activity permits industry and government scientists to work together in laboratories, completely sharing information if that will foster the aims of the research, which must be of common interest to both entities. In a very real sense it provides an extension of a company's laboratory capability for those interests that closely parallel the interests of the cooperating government lab.

A third incentive for industry to participate is the substantial opportunity to reduce R & D costs through access to reliable, high-quality technology. This access may be to reports, research scientists, unique equipment or facilities, or some cooperative venture. At the very least it can provide the type of information and experience by which a company can avoid duplicating past mistakes or reinventing information, while at the best it can exploit recent breakthroughs, turning them to commercial purpose. Beyond this type of cost reduction, there is the reduction through cooperative endeavors where there is, in fact, a true sharing of costs. Here industry has the advantage of access to *all* of the scientific results while only having to pay a portion of the total cost.

A fourth incentive to industry is the opportunity to use unique facilities, have access to quality researchers, and to an extensive store of knowledge that is not otherwise available. The purpose for such access may be problem solving, or it may be the kind of stimulation that produces innovation and new products through technical synergism. Federal laboratories have a significant array of unique facilities unavailable to private industry in other circumstances. For decades organizations such as the National Advisory Committee on Aeronautics (now NASA) and the Atomic Energy Commission (now the Department of Energy) permitted industrial counterparts to make use of their unique laboratory facilities. Now this opportunity is extended by all federal laboratories.

Cooperation of scientists from industry, government, and often universities, can have a valuable synergistic effect. For example, the programmable implantable medicinal supply (PIMS) device resulted from a joint effort by Goddard Space Flight Center, the Applied Physics Laboratory of Johns Hopkins University, and several medical device manufacturers. The device provides in a small package an implantable pump with reservoir—in recent clinical trials the principal element has been insulin for treating diabetics—by which the medication can be supplied automatically to the body as needed, with adjustments made by telemetry from an external source. This gives substantial freedom of action for the person needing the medication, provides a medically more secure and accurate means of medication, and frees the patient from the often painful (to say nothing of derivative) side effects of numerous injections. The PIMS system was a result of cooperative technical synergism.

A fifth incentive for industry consists of substantially broadened new licensing opportunities. Before the recent legislative changes, it was rare for a government agency to issue exclusive licenses—thereby requiring the industry to take the risk that competitors might easily enter the field after an investment in the development costs to commercialization had been completed. Now, government agencies can enter CRADAs early in the course of development and can use exclusive licensing to commercialize the technology that results from the cooperative research. This approach is becoming very nearly standard practice in many of the CRADAs initiated today. However, one should not view this as a risk-free way to obtain an exclusive license. In most instances such exclusivity flows only from truly cooperative efforts in which industry has invested significant resources.

INCENTIVES FOR GOVERNMENT

Evidence of an agency's active participation through use of CRADAs and provision of technical assistance and advice to meet problems in U.S. industry, leads to favorable consideration of agency R & D programs in the congressional budget decision process. Accounts of successful transfers that helped U.S. business, solved a local government's problem, or added to local employment help agencies secure funding. Of course, these examples alone are not sufficient to justify substantial R & D programs. On the other hand, demonstrating that these programs serve several purposes, including contributions of value and benefit to local constituencies, provides an advantage in the budget process. Testimony from pleased users of federal technology or examples of spinoffs are especially powerful.

Technology transfer is a two-way street. Although recent legislation provides industry access to federally developed technology, cooperative arrangements by which industrial scientists and federal laboratory scientists come into contact benefit government agencies. The technology transfer process provides government access to industry's technology and, at the very least can result in broadened and enhanced technological insight on the part of the government participants. There are numerous examples of technological *spinback*—the phenomenon by which provision of technical assistance to an outside party generates novel technical solutions, which, in turn, result in significant value to the assisting laboratory. For example, some years ago a nondestructive evaluation technology developed within NASA to detect strain in bolts in a wind tunnel application, eventually was used in a device for mine safety purposes. NASA, cooperatively with the Bureau of Mines and the mining industry, developed a relatively simple instrument that could detect strain in the bolts that secured the ceilings in underground mines. This technological advance developed for uses external to NASA in turn proved valuable when NASA had to diagnose a similar problem in the wheels of the space shuttle.

A third incentive for government cooperation is through CRADAs, which provide government laboratories access to a hitherto difficult or denied source of external funding. Before this authority, industry co-funding for R & D involving a government partner was rare. This authority allows agencies to accept supplemental resources to further research of interest both to the industry supplying the funds and to the laboratory undertaking the work. This then serves the double purpose of meeting industry needs as well as assisting the work of the government laboratory.

Finally, both laboratories and individual government scientists benefit by sharing royalties from license income. The Technology Transfer Act of 1986 required that government scientists responsible for a licensed invention will be given no less than 15 percent of the royalties that result from such license. Most agencies in the federal government have increased this to 20 or 25 percent, and often the inventor will receive much of the first $2,500 or $5,000. Perhaps even more important than this sharing with the individual inventors, much of the remaining license income is allocated to the laboratory where the invention was made. This permits the laboratory director to supplement the research budget of the organization that produced the invention, as well as to facilitate further technology transfer and education activities of the laboratory. No longer does the inventing organization lose the money to the U.S. Treasury.

Significant incentives indeed exist for both industry and government to seek out maximum cooperation in R & D endeavors, searching for those circumstances in which technological common ground can result in substantial technical progress.

CHAPTER 5

HOW ARE THE AGENCIES ORGANIZING TO RESPOND? AGENCY PERSPECTIVES AND MANAGEMENT SUPPORT

This chapter provides a thumbnail sketch of the technology transfer infrastructure and policies of the most research-oriented federal agencies. These consist of eight departments: the U.S. Departments of Agriculture, Commerce, Defense (DOD), Energy (DOE), Health and Human Services, Interior, Transportation (DOT), and Veterans Affairs (VA), and two independent agencies, the Environmental Protection Agency (EPA) and NASA.

Policies at the departmental level for guiding the transfer of technology to domestic use have developed slowly. Perhaps the best known is that of the Agricultural Extension Service, the roots of which go back more than a century. However, in terms of the transfer of technology from federal laboratories, the first clear and explicit policy was established in 1963 by NASA when it initiated its agencywide Technology Utilization Program.

Only since the passage of the Technology Transfer Act of 1986 has there been much serious guidance at the departmental or independent agency level. Some of these policies still are seldom implemented, but virtually every agency has at least some policy guidance in place. Sometimes there is initiative at the bureau level. In nearly every instance there is considerable activity at the field laboratory level—some of it well in advance of the guidance provided either through their respective departmental/agency headquarters or from their bureaus. One can anticipate considerable progress in the future, providing that Congress continues its support for this closer cooperation with industry, and provided that the departmental or agency leadership appreciation of the value of technology transfer continues to increase.

AGRICULTURE

The Department of Agriculture has two primary research organizations conducting in-house research: the Agricultural Research Service (ARS) and the research arm of the U.S. Forest Service. Their combined research budgets exceed $800 million annually. Each has a formal point of contact for facilitating inquiries from outside the department concerning technological opportunities—both informal advice and licensing, as well as the opportunities for cooperative activities.

The National Agricultural Library provides information about technology transfer activities within the department. A similar contact point is located within the national headquarters of the Agricultural Extension Service, whose federal-state-county network extends throughout the United States. A second research and technology network, the Cooperative State Research Service (CSRS), is responsible for a major research grant program that provides funds to the land grant universities in each state, principally through their individual experiment stations.

Agriculture has a decentralized system for technology transfer. Although the assistant secretary for science and education was given the primary responsibility as the principal official within the department for general contact, the responsibility has been delegated to the ARS and the Forest Service. Each of their major laboratories or research stations throughout the country has a contact point for persons or organizations seeking technical assistance. Go directly to the individual laboratories for assistance, though sometimes it may be more fruitful to go to the principal agency contact who has an overview of all of the technical activities within the agency.

The Department of Agriculture moved rapidly in 1987 to implement the Technology Transfer Act of 1986. Authority for conducting technology transfer and particularly for entering into the cooperative research and development agreements (CRADAs) was quickly delegated from the secretary to the administrator of the ARS and the chief of the Forest Service, respectively. Each of these two agencies has taken a somewhat different path. Both have undertaken an aggressive stance towards patent licensing.

Agricultural Research Service

The ARS has provided a central point of guidance and activity for technology transfer across the service, using that central point for final clearance and signoff for CRADAs. In order to facilitate CRADAs, ARS published the pamphlet, "Technology Transfer Agreements between Industry and ARS: A Plain Language Guide." ARS also has distributed a videotape to the field providing an orientation to stimulate its bench scientists. ARS has implemented more than one hundred CRADAs with nearly three dozen patents awarded and royalty income of over $400,000.

Since 1989 ARS has made annual awards for outstanding contributions to technology transfer. The service makes available a current, on-line data base of ARS research projects called TEKTRAN, accessible directly from a user's personal computer.

Forest Service

The Forest Service has placed its ORTA with the deputy chief for state and private forestry, but it has delegated the responsibility for CRADA signoff to its principal field laboratories. The Forest Service has developed an on-line data base for inquiries. The Forest Products Laboratory, in cooperation with the state and private forestry organizations, publishes "Technology Opportunities," which provides summaries of Forest Service technology for applications in the form of one- and two-page summaries titled "Tech Line." The Forest Service had as many as twenty-nine nominees for its annual technology transfer awards, which can amount to $10,000 each—one of the largest award programs for technology transfer activity.

National Agricultural Library

The National Agricultural Library recently established a Technology Transfer Information Center for the use of the department and its clientele. It is in the process of putting together a technology transfer collection of documents and research papers, and will provide a point of reference within the department for departmental technology transfer activity.

Commerce

Like other departments and agencies, the Department of Commerce provides guidance to its internal constituent elements. Commerce also has responsibility under both the Stevenson-Wydler Technology Innovation Act (1980) and the Technology Transfer Act of 1986 to provide governmentwide assistance in technology transfer. At the secretarial level there is an undersecretary for technology who heads the Technology Administration. This administration has oversight of the National Institute of Standards and Technology (NIST), formerly the National Bureau of Standards, the National Technical Information Service (NTIS), and the National Telecommunications and Information Administration (NTIA). The undersecretary also houses a newly established assistant secretary for technology policy wherein resides oversight for technology policy, technology commercialization, and the Clearinghouse for State and Local Initiatives on Productivity, Technology, and Innovation.

This policy staff assists the other agencies by chairing an interagency group at both the assistant secretary and working levels. Common problems are discussed, including the development of CRADAs, defining appropriate relationships and working with foreign corporations, questions on conflict of interest, patent-licensing problems, and the development of common training programs. The assistant secretary provides a focal point for the commerce laboratory directors to deal with particular commerce-oriented questions of technology transfer.

The Department of Commerce has three long-standing agencies that have been involved in domestic technology transfer. These are NIST, the National Oceanographic and Atmospheric Administration (NOAA), and NTIS. The first two have extensive laboratory facilities; the third, NTIS, is essentially a service organization that provides reference to documentary material deposited with it from all federal departments and agencies. The NTIS can provide a variety of information services —both printed and through computer disks or on-line access.

NATIONAL INSTITUTE OF STANDARDS AND TECHNOLOGY

NIST recently established a technology services group that coordinates a variety of activities, including the classic technology transfer activity of previous years. It includes the following five elements: (1) Manufacturing Technology Centers Program, (2) Inventions Evaluation Program, (3) Measurement Services, (4) Standards Services, and (5) Technology Commercialization.

NIST also manages two recent programs to accelerate technology transfer in U.S. industry. One development and demonstration program uses three manufacturing technology centers to assist small and medium businesses to upgrade their manufacturing technology and thereby become more competitive. The second, the Advanced Technology Program, helps U.S. business conduct R & D, through cooperative efforts, on precompetitive, generic technologies. Both programs aim to facilitate technology transfer more broadly by research and demonstration through joint government-industry efforts.

National Oceanographic and Atmospheric Administration

The Office of Research and Technology Applications (ORTA) function is now located under the direct authority of the chief scientist, giving it more visibility. NOAA publishes a research and development guide, a survey of NOAA technology available for transfer, and continues the publication of summaries of NOAA technology titled *NOAA Tech Briefs*. Despite the relatively narrow technological base of the NOAA laboratories, there is an increased interest in CRADAs and a marked increase in invention disclosures.

Defense

The Technology Transfer Acts of both 1980 and 1986 treated the military departments (U.S. Army, Navy, and Air Force) as independent departments. Therefore, minimum guidance is provided at the office of the secretary level. DOD Regulation 3200.12-R-4 was issued in December 1988 delegating the responsibility for implementation of the Technology Transfer Act of 1986 to the military departments and giving them the authority to enter into CRADAs.

Army

A dozen or more army laboratories have been active in technology transfer for well over 10 years. The U.S. Army was the first military department to have a regulation providing guidance and stimulation to technology transfer, even pre-dating the 1986 act. The lead was taken by the Army Material Command, and now by the Army Laboratory Command (LABCOM).

Army leadership is pressing to establish full-time ORTAs at all of the army's major laboratories. The ORTA function has been institutionalized at some 42 army labs. The ORTAs are being encouraged to take the lead in resolving potential conflicts of interest at an early stage, acting as a third party in representing the laboratory while providing assistance to the laboratory scientists.

There is strong interest within the army to use the Small Business Innovation Research (SBIR) program as an outreach activity to encourage further technology transfer—especially in the phase II and phase III stages, or those leading to commercialization. The army has moved rapidly in the past two years to facilitate interaction with industry through CRADAS.

Navy

The navy regulation implementing the Technology Transfer Act provides delegation of the CRADA authority from the secretary of the navy to the chief of naval research. It also outlined general guidance on technology transfer. The U.S. Navy has instituted a fast track for CRADA approval that is being retained by the chief of naval research.

The Office of Naval Technology ORTA has produced videos for orientation of both navy and industry lab scientists. The U.S. Navy continues to produce its *Domestic Technology Transfer Fact Sheet*, which presents technology having potential industrial application. These notices are now being featured in *Aerospace and Defense Science Magazine*. The navy has participated in several technology transfer fairs where navy technology is highlighted, as well as cooperative ventures with other organizations such as NASA.

Air Force

The U.S. Air Force issued its implementing regulations (Air Force Regulation 80-27, Domestic Technology Transfer) in January 1990, which delegated the CRADA responsibility to the subordinate commands or laboratories, and provided extensive guidance for the technology transfer function. This guidance incorporated domestic technology transfer into each laboratory's mission, made it the responsibility of each laboratory scientist and engineer to participate in technology transfer activities, and directed that such activity should be included in the job evaluations of these individuals. ORTAs have been established at all laboratories. The air force headquarters is introducing the technology transfer requirement into position descriptions.

Strategic Defense Initiative Organization

One of the earliest organized efforts with specific funding for technology transfer within DOD was the Strategic Defense Initiative Organization (SDIO). SDIO established a Technology Applications Office which, in turn, has developed the Technology Applications Information System (TAIS) with more than 13,000 nonclassified items of technology available for potential commercial use. In addition, SDIO has proceeded, under a congressional authorization, to establish five regional Medical Free Electron Laser Centers, which involve cooperation with universities, industry, several federal national laboratories, and teaching hospitals.

Energy

Beginning in January 1990, DOE undertook a major policy review to determine the best way to carry out aggressive technology transfer to industry, and to make it a key part of national energy policy. This activity, conducted by field and headquarters task groups, both of which completed several policy developments and reviews, resulted in several promising initiatives. It involved for the first time the defense-related programs, so that the *entire* department has guidance, policy, and management support for more aggressive technology transfer. Second, the secretary of the department, as well as policy subordinates, has clearly and strongly demonstrated support of future technology transfer activities within the department.

As noted earlier, P.L. 101-189 extended the National Competitiveness Technology Transfer Act of 1989 to government-owned, contractor-operated (GOCO) laboratories—specifically to include the large national laboratories owned by DOE.

Health and Human Services

The two primary research organizations within this department are the National Institutes of Health (NIH) and the Food and Drug Administration (FDA). Much of the policy and activity has been delegated to these two bureau-level agencies within the department.

National Institutes of Health

Taking the lead for the department, NIH has been joined by the Alcohol, Drug, and Mental Health Administration (ADAMHA) and the Centers for Disease Control (CDC) in a joint effort as part of the implementation of the Technical Transfer Act of 1986. NIH established the Office of Technology Transfer (OTT) under the associate director for intramural affairs. Each of the institutes, centers, or divisions has a technology development coordinator who acts as the principal liaison and action point for each of these elements and works with OTT and with the NIH Patent Policy Board.

NIH conducts an annual industry collaboration forum at which technologies available for collaboration with industry are showcased for NIH, ADAMHA, and CDC. The NIH recently joined with the Pharmaceutical Manufacturers Association to conduct a two-day national technology transfer conference. More than two hundred CRADAs are in place, and others are being developed and negotiated.

NIH has taken the lead in examining ways to prevent potential conflict of interest by conducting a two-day workshop of its senior executives.

FOOD AND DRUG ADMINISTRATION

An ORTA has been established at a senior level within the office of the commissioner. A model CRADA is in use, with several agreements currently in place and more in development. As primarily a regulatory agency, the FDA is carefully developing its outreach program, reviewing concerns about potential conflict-of-interest issues, and orienting FDA staff.

INTERIOR

The Department of the Interior has four major bureaus involved in the implementation of the P.L. 99-502 1986 Technology Transfer Act. They are: the Bureau of Reclamation, the Fish and Wildlife Service, the Bureau of Mines, and the U.S. Geological Survey (USGS). The department essentially has delegated authority for implementation of the act to its bureau-level organizations without extensive guidelines.

BUREAU OF RECLAMATION

The Bureau of Reclamation established an ORTA for the bureau as a whole, but it delegated the CRADA authority to the bureau's laboratory directors. The bureau recently issued a reclamation instruction on technology transfer. Bureau officials have seen an increase in invention disclosures and have begun the process of negotiating licenses for technologies that have been patented, as well as the implementation of CRADAs. The bureau is taking the lead in examining cooperative research among universities, institutes, other agencies, and private industry for water concerns in the western United States.

FISH AND WILDLIFE SERVICE

The Fish and Wildlife Service has had an ORTA function that considerably predates the 1986 act. It has been located in the extension group of the research organization within the service. They work primarily with fish and wildlife organizations in the states and therefore do not often relate directly to industry, dealing more with the exchange of scientific information with state organizations and with universities. The service updated its administrative manual to reflect its responsibilities under the Technology Transfer Act of 1986, and it has established contact points in each of the service laboratories around the United States.

BUREAU OF MINES

The bureau continues a long-standing technology activity through its technology transfer branch in its headquarters in Washington, DC. Here, they oversee the publication of technology summaries and foster both demonstration and workshop activity for the transfer of technology to the mining industry. The bureau has more than two hundred cooperative agreements involving testing or demonstrations. It has explored the need to fill the gap between research and industrial development, and it is surveying all federal labs for technology appropriate to the mineral industry. Three labs, located in Pittsburgh, Birmingham, and Denver, have technology transfer offices.

U.S. GEOLOGICAL SURVEY

An ORTA was established at the level of the assistant director for programs in USGS headquarters in Washington. This officer coordinates the activities of USGS field liaison officers and is a focal point for inquiries.

TRANSPORTATION

The Department of Transportation (DOT) consolidated its responsibilities for technology transfer under the secretary within its Office of Research Policy and Technology Transfer. This office retains oversight of the operational activities involving technology transfer in the Urban Mass Transit Administration (UMTA) and the Federal Highway Administration (FHWA). UMTA continues Project Share, which is the provision of technical assistance and technical advice to the department's principal constituencies comprising mass transit organizations in cities and the state and local highway organizations.

Within the FHWA there has been a substantial reorganization to pull together its Office of Technology Applications (responsible for the implementation function), elements of the National Highway Institute (primarily directed to rural highways), and demonstration projects. This will provide a more effective means for the transfer of technology and useful "hands on" demonstrations. Because nearly all of FHWA's technology transfer involves direct relationship with government organizations at the state and local levels, there has been little interest in CRADA type activity of a commercial nature.

VETERANS AFFAIRS

Departmental headquarters published a circular letter in December 1989 to all of its hospitals and centers, providing general guidance and indicating the process by which CRADAs would be developed. Several CRADAs have been developed. That process requires joint review by its Office of Research and the general counsel's office. There has been a notable increase in invention disclosures. The department has a policy unlike other federal agencies in that almost all VA inventors are permitted to keep title to their inventions and pursue commercialization.

ENVIRONMENTAL PROTECTION AGENCY

EPA recently established the National Environmental Technology Applications Corporation (NETAC) at the University of Pittsburgh. NETAC acts both as a broker and facilitator of technology development to bring it to commercialization —primarily in conjunction with corporate or other partners. NETAC matches funds provided by EPA. NETAC is establishing a series of testing protocols to determine what oil spill remediation techniques are workable and safe. NETAC also conducts market surveys and analyses relating to the mix of federal and state regulations and the identification of technologies that are safe and effective for hazardous waste cleanup.

EPA recently established the Alternative Treatment Technology Information Center (ATTIC), which provides information about treatment technologies in a single data base for more than eighty such technologies. The *Engineering News Record* recently gave an award recognizing EPA officials for this accomplishment.

NATIONAL AERONAUTICS AND SPACE ADMINISTRATION

NASA, which has the oldest formal technology transfer activity, the Technology Utilization Program, opted to use its flexible authority under the National Aeronautics and Space Act of 1958 (as amended) in lieu of the Technology Transfer Act of 1986 to carry on its technology transfer activities involving spinoff activity. NASA has provided guidance for patent-licensing requirements and for sharing royalty revenue among inventors. It also delegated CRADA authority to its field center directors. NASA recently showcased NASA technology having potential for commercialization at an annual conference called "Technology 2000."

The special publication of NASA technology, *NASA Tech Briefs*, has been published commercially for several years and now has a circulation of approximately two hundred thousand monthly. Benefits from only 259 applications of NASA-furnished technology contributed to sales or savings of nearly $22 billion, additional federal corporate income taxes of $356 million, and creation or saving of 353,000 jobs.

This brief overview of agency-level responses to the mandates of recent technology transfer legislation indicates commitment that is being translated into action at the laboratory level.

CHAPTER 6

THE LABORATORIES

The laboratories where in-house research is conducted are the heart of the federal R & D system. These laboratories vary widely in mission, funding, size, and characterization. If each research location of the federal government is called a "laboratory," then there are more than six hundred. However, many are field stations with only a handful of scientists or engineers conducting field experiments. They can range in size from a handful to more than five thousand technical personnel in a single location working in large teams on a variety of projects.

THE BIG NATIONAL LABORATORIES

The largest research laboratories or research centers are run by the Department of Defense (DOD), the Department of Energy (DOE), and NASA. DOD has more than seventy major labs, and their total laboratory system employs approximately sixty thousand scientists and engineers. DOE runs twenty-one major multipurpose or single-purpose laboratories, and has ten other lesser laboratories —the total system employing approximately twenty-two thousand scientists and engineers. NASA has nine major research or flight centers employing approximately ten thousand scientists and engineers.

Each of these large national labs has a primary in-house mission—usually to support the functions of the department that sponsors it. In the case of DOD it is national security, and its primary clientele will be the men and women of the armed forces. In the case of the DOE, a portion of its laboratories (such as the weapons labs) will be devoted to the national security mission as are the DOD labs. However, DOE also has clientele across U.S. industry related to nuclear, solar, mineral-based and other types of power, and the necessary safeguards such power requires. NASA serves as a research arm for aeronautics—both government and the private sector—as well as for its own peculiar aerospace missions and general support for the aerospace industry.

These missions of the major laboratories are not necessarily narrow. To carry out any of these missions, including energy, national security, and aerospace, it is necessary to explore a tremendous variety of scientific and technological possibilities, ranging from the physical and psychological role of the human being to the production and use of materials, electronics, computer science, and mathematics. Virtually any of these technologies have potential use outside the mainstream mission of the respective laboratories.

OTHER FEDERAL LABORATORIES

The other federal departments and agencies that have an important technological base typically support considerably smaller, although certainly no less important, laboratories. These include the Departments of Agriculture, Commerce, Health and Human Services, Interior, Transportation, Veterans Affairs, the Environmental Protection Agency (EPA), and the National Science Foundation (NSF). For example, the Agricultural Research Service (ARS), which receives funding of approximately $700 million per year, has research at more than fifty locations across the United States, led by four regional research centers. But this group of laboratories is closely interconnected with the research experiment stations run by the land grant universities throughout the United States. Indeed, broad research plans tend to be integrated, and scientists from ARS and state experiment stations often work together in the same labs.

The Forest Service has about eight hundred scientists and engineers in their research organizations, which are located at seven principal regions in more than fifty research locations. They, too, relate closely to land grant institutions and schools of forestry.

Commerce's two principal technical agencies, the National Oceanographic and Atmospheric Agency (NOAA) and the National Institute of Standards and Technology (NIST), serve more general clientele. NOAA provides information worldwide of great importance to the general public, as well as to transportation organizations and any industry affected by weather and the oceans. On the other hand, NIST has the key function of establishing measurement standards that are critical to commerce throughout the United States and the world, as well as providing general technical backup to U.S. industry in this broad area.

The principal technical agencies in Health and Human Services are the National Institutes of Health (NIH), under the Public Health Service, and the Food and Drug Administration (FDA). They serve both the general public through the health care professions and also organizations that provide health care products or services. The NIH intramural laboratories, located principally in Bethesda, Maryland, focus on fundamental and clinical research and are considered the premier medical research organizations in the world.

Among Interior's laboratories is one operated by the Bureau of Reclamation that focuses on all aspects of the construction and maintenance of irrigation and hydroelectric systems, and studies a whole array of environmental factors. The Bureau of Mines conducts research at three principal locations on mine safety, development and maintenance of equipment, and improved techniques for economic recovery of minerals. Both bureaus work closely with universities or colleges where the sciences related to mining are taught. The Fish and Wildlife Service of the Department of the Interior has research field stations in each of its regions and works closely with its state counterparts as well as with the Forest Service and the National Park Service.

DOT has a Transportation Research Center in Cambridge, Massachusetts, several Federal Aviation Agency research locations, and a transportation testing station in Colorado.

The VA generally has a research function attached to each of its major hospitals, which typically are located adjacent to medical schools or similar research organizations across the country. Some hospitals specialize—for example, some locations are deeply involved in rehabilitation research.

The EPA has several primary laboratories involved in environmental research, but it has a substantial number of laboratories principally devoted to testing for regulatory purposes rather than conducting research. Again, although EPA's clientele ultimately is the general population, its scientists work most closely with compatriot organizations in state and local governments, and with industrial groups where key environmental problems have been identified.

The NSF sponsors several major laboratories such as the radio telescope in Greenbank, West Virginia, and the National Center for Atmospheric Research in Boulder, Colorado. These principally conduct fundamental research for supporting science in academia and industry.

Organization and Environment for Technology Transfer

The act that originally established technology transfer as a function for all agencies conducting research and development was the Stevenson-Wydler Technology Innovation Act of 1980. This act provided for an Office of Research and Technology Applications (ORTA) that was to be established in each federal laboratory with an annual budget of $20 million or more. Although agencies were slow in following this guidance, most now have such a function designated and staffed at least part time. During the past decade this position increasingly has been occupied by individuals who have a personal interest in the technology transfer function. Indeed, the bulk of the Federal Laboratory Consortium (FLC) membership is made up of the ORTAs. Generally, the scientist or engineer who holds the ORTA position is an individual who has a broad understanding of current research activities across the particular laboratory. In a number of instances the ORTA may also have parallel responsibilities for handling technical information and the Small Business Innovation Research (SBIR) programs—all of which add to the scientist's broad knowledge of current activities.

The purpose of the 1980 legislation was to establish a central point of contact for persons outside the laboratory to link with technological opportunities within the laboratory. The ORTA generally fulfills this function. However, as is often the case, individuals outside the laboratory who are deeply involved in a particularly technological area may have personal contacts within a laboratory and link through that source rather than through the ORTA.

Ultimately, you want to make contact with a lab scientist who has the experience and knowledge you seek. Most scientists are inveterate problem solvers and also enjoy discussing their current research projects with virtually any other individual who can understand the technology and its importance. Anyone making a technical inquiry, particularly if the individual seems to be technically competent to discuss the technology involved, is likely to be well received.

However, there are some exceptions. Necessarily, scientists are sometimes deeply absorbed in the conduct of an experiment and will not tolerate interruption. There also will be periods of administrative stress, such as program or budget reviews, during which a scientist is likely to be less responsive. Other factors that affect responsiveness of lab scientists include the nature of the laboratory and the general stance of the scientist's immediate laboratory supervisor.

There are considerable differences from one laboratory to another. For example, a research center or laboratory is likely to be under less time stress than one involved in applications engineering, weapons production, or flight activities. The time stress involved in major projects often provides a psychological atmosphere in which the scientist or engineer avoids external distraction. This is not universally true, but you should expect to find that attitude occasionally among scientists and engineers at research laboratories or centers where large applications projects are in progress.

The other primary factor is the attitude of the scientist's immediate laboratory supervisor. If that supervisor is open to external problem-solving activity, the opportunities for responsiveness are considerably enhanced. There have been supervisors in federal laboratories who were absolutely opposed to such activity, taking the attitude, "If you have time for that, I question your usefulness to the research team." This certainly is not the norm. However, you should not be overly discouraged by such an attitude. Usually, there are ways to go around such barriers, for instance, by soliciting the help of the ORTA or a higher-ranking authority within the laboratory structure.

Finally, the laboratory management tends to set the tone for the entire organization. The laboratory or center director can establish an environment in which external problem solving and cooperation with those from outside (other than their contractors and immediate collaborators) are encouraged. There has been considerable improvement over the past decade in senior management's toleration and even encouragement of technology transfer activities. On reflection, most managers recognize that the laboratory gains as well as does the particular client in such exchanges. The scientists or engineers involved typically expand their technical horizons, find considerable pleasure in such activities, and thereby are sharpened for their primary mission within the laboratory.

Even where laboratory management does not seem particularly receptive to technology transfer activities, there can be considerable activity of this nature conducted at the branch and bench level where the participants are most willing (government employees are now generally aware they will benefit from the required royalty sharing). This often may go unnoticed by management simply because the visibility of such branch- and bench-level activities often is not very high. However, as Congress and executive agencies have put greater emphasis on technology transfer activities in terms of being helpful for the nation as a whole, there has been an increasing interest and awareness on the part of laboratory directors.

SUCCESSFUL ENTRY TO THE LABORATORY

The most obvious first point of entry to try is through ORTA if there is one at the particular laboratory. If there is no such organization, or if the position is temporarily vacant, you can next go to the coordinator of the FLC in that region. He or she generally will have points of contact in most laboratories, even where there is not an ORTA function. Another possible point of contact is through the agency representative to the FLC. Each agency has such a representative, and that individual usually is conversant with the capabilities (in a general sense) at each agency laboratory and can guide you to an initial contact in the laboratory of choice.

If all else fails, you should not neglect using a technology transfer broker. This could be a Regional Technology Transfer Center (RTTC) sponsored by NASA, an agency-operated technology information office, or even a private company or entrepreneur who acts as a technology transfer broker. For example, NERAC, formerly a NASA Industrial Applications Center, has a special program, "Expert Match." A person making the inquiry will discuss in detail with NERAC specialists the nature of the technology required. NERAC then, using a list of experts in the federal laboratories, will contact a liaison in the laboratory where one or more of the experts is located. Once NERAC has made a satisfactory contact with an expert to assure that the client can talk directly to that individual, the client is thus directed, and the link is completed. This process can be particularly useful where there is some other institutional blockage to a knowledgeable inquiry.

There are so many federal laboratories and research locations, that you should be persistent in searching for technical help or technical opportunities. A single laboratory is unlikely to have a monopoly on any particular technology sought—there are probably several sources for whatever you are seeking.

CHAPTER 7

YES, BUT WHAT ABOUT...?
THE MYTHS AND REALITIES OF
TECHNOLOGY TRANSFER

The opportunities for significant gain through participating in the transfer of technology from federally funded R & D are great. However, you must be wary, recognizing the difficulties and challenges as well as the significant rewards. There are no guarantees in going from innovation to product. For example, NASA and a team of outside collaborators have spent years and many millions of dollars—and even overcome FDA hurdles—in developing the programmable implantable medicinal supply (PIMS) for sufferers of diabetes, yet its acceptance may be stymied if health insurers do not agree to cover the costs of these devices. This experience should not dampen your enthusiasm, but it should provide a necessary backdrop of caution against which to undertake truly fruitful planning and action.

AVOIDING THE MYTHS

Several years ago, an editorial appearing in TRW's *Electronics and Defense/Quest* (an in-house magazine) pleaded the case for working at technology transfer by combating several misconceptions. Common myths are that: (1) industry automatically devours new technology as soon as it is revealed; (2) a better mousetrap is self-evident and does not need selling; and (3) so-called exciting and valid technology will *automatically* be transferred. All of these myths are founded on the erroneous belief that worthwhile transfer is a self-servicing system. The basic point of the editorial was that a company or agency must be organized to enhance technology transfer if the right connections are to be made and if technology is to be applied most effectively. And this requires deliberate effort throughout an organization.

The federal lab employee is certain that he is doing world-class work and is not prepared for the indifference that is sometimes encountered on the outside. A good example is a Department of Energy effort to interest companies in exploiting their $15 million synchroton R & D that could be further developed for X-ray lithography. Though opportunities for cooperative research and development agreements (CRADAs) were published in both the *Commerce Business Daily* and *Federal Register*, no one responded. The people involved in the federal labs are awakening to the tough job of selling a new technology.

Another frequently accepted misperception is that technology transfer consists primarily of accessing data bases through written publications or electronic means to dig out nuggets of technology. This is an excellent first step to screen a wide variety of potential opportunities or problem solutions. Typically, however, it is only a first step and rarely will produce an instantaneous solution. William D. Carey, former executive officer of the American Association for the Advancement of Science, noted in a letter to one of the authors, "Technology transfer is a hands-on process and not a mail room or brochure-type activity."

TECHNOLOGY TRANSFER REALITIES

Nearly three dozen corporate executives were interviewed concerning their perspectives on technology transfer. These executives represented a wide variety of businesses including high technology (electronics, aerospace, and medical equipment), automobile and heavy manufacturing, engineering and product development, and investment capital. Both large and small firms were represented. The discussion included such topics as changes in the pattern of technology acquisition over the past twenty years, influences of foreign competition, channels of technology transfer, industrial patent policy, university linkages, and barriers or incentives to the transfer of technology.

The structural differences between government and industry were pronounced. One is politically responsible and the other is responsive to market forces. Even though these differences tend to generate skepticism among industrial managers, there remained a strong interest in the potential for cooperation *and* a realization that much can be gained through improved transfer or exchange of technology.

These industrial managers generally believe government-generated technology includes many innovations that are or can be attractive to industry. On the other hand, they felt that government technology—especially high technology—might be too expensive for commercial application without substantial adaptation. Industrial leaders stress that this adaptation—in many cases, redesign and engineering costs—often may reduce the original estimate of the value of particular innovations.

A general concern among industrial leaders seems to be that federal managers are not sufficiently aware of the types of constraints with which industry must deal. These include a competitive economic environment, problems with marketing, and the protection of proprietary information.

Patents are often essential to a company's decision to invest development and marketing capital in a new technology, yet they have little or no bearing on the success of a particular federal lab's mission. As a result, there is great suspicion on the part of industry that federal employees are unaware of and not diligent in observing the essential inventing practices of documenting their work and of making timely disclosures. Information made available is often directed strictly to a technical audience without regard for the concerns of the entrepreneur-protected production costs and the technology's application to other fields or problems. If offered to everyone without targeting selected companies sequentially, the research becomes less attractive because it is impossible to launch a new product that catches competitors off guard.

Additional concerns arise when it becomes known that even though a CRADA relationship can include up-front licensing of CRADA-developed inventions, prior patents that may be essential to a good intellectual property package cannot be guaranteed to the collaborator and must be advertised in the *Federal Register* if they are to be licensed exclusively. Businesses often question the rights reserved for the government to "make or have made" the invention for government purposes, as to whether this will interfere with their ability to market the technology to the government. Finally, how willing is the government to pursue patent infringers? The small number of federal patent suits makes it appear that the stance is not very aggressive. In considering major investments to commercialize a new technology, a prospective licensee may be dissuaded by the prospects of having to fight a larger firm on its own. Fortunately, the government is aware of these concerns and the Commerce Department is, in fact, holding public hearings to recommend changes, if needed.

When the focus is on federal-employee-created software, which as of today cannot be copyrighted, U.S. businesses feel they are at a distinct disadvantage. Indeed, a recent court case negated a company's copyright on one of their products, because it was ruled that the basic software had been developed at a federal lab. See Chapter 12.

One sure way to make technology transfer discussions more realistic is to mention the Freedom of Information Act. Fortunately, there are guidelines for protecting proprietary data, but both sides must be aware of and diligent to comply with them. Sensitive information that a company already owns should clearly be marked as such, and its inclusion should be kept to a minimum. Data relating to patentable inventions arising during a CRADA or contract should be described and evidence of pursuit of patenting should be provided. Some business-sensitive information developed in a CRADA can be kept secret for five years. Otherwise it is public domain.

Some of the people interviewed want clearer guidelines on important areas left open to interpretation. The enabling legislation favors U.S. companies and small businesses, but today, just what is a U.S. business? If one cannot be found, a company with "substantial manufacturing" in the states is the next choice, and then those whose countries have reciprocity in trade. Owners of small businesses may be faced with unofficial stonewalling because of the local tech transfer official's lack of confidence in their ability to perform. More official reluctance results when DOD is unwilling to decide if someone with more authority has not made a determination on a technology's importance to national security. Further, what happens when you are working with a government-owned, contractor-operated (GOCO) lab that has a private company as an operator and the management is replaced with a nonprofit, or vice versa?

If the technology being developed will require regulatory acceptance, such as by the FDA or Environmental Protection Agency (EPA), the collaborating business may become wary of the government partner's willingness and ability to perform in the long term. Each year the possibility of budget cuts by Congress means that the lab's commitment to the partnership is tentative.

A common frustration mentioned by both companies and civil servants eager to see their work commercialized is the slowness inherent in dealing with the government bureaucracy. Many reasons for this inertia have been illustrated humorously and not so humorously by social commentators for as long as there have been governments. A report published by the General Accounting Office (GAO) in 1990 concerning the implementation of the federal Technology Transfer Act of 1986 (P.L. 99-502) reflects this slowness. The report, though preliminary, shows that although some progress has been made, there is still much left to be done to fully implement the federal side of technology transfer.

Of the 187 federal laboratories surveyed, only 110 had received final written instructions from their agencies for implementing any or all parts of the act; 29 had received draft instructions, and 48 had received no instructions at all.

Few of the labs (37) had the Office of Research and Technology Applications (ORTA) that they controlled actually within their laboratories. Of the others, 128 ORTAs were at their controlling agency headquarters and 16 were at other locations. Additionally, it must be remembered that it takes time for changes in response to new legislation to become familiar, as new operating policy, to several different departments—ORTAs, legal, contracting, and R & D management—each with their own bureaucratic inertia.

Disappointingly, especially four years after the passage of the act, only 71 of the labs had received authorization from their agencies for approving CRADAs with outside industry. Even more disappointing is the fact that only 34 of the labs had a representative to the Federal Laboratory Consortium (FLC).

Most labs placed very little emphasis on staff involvement in technology transfer. Only 41 of the labs gave any type of award (separate and distinct from those given by their agencies) to reward scientific, engineering, and technical personnel for activities leading to the filing of patent applications or the award of patents. Even though it is specifically mentioned in the law, only 41 of the labs had any guidelines that specifically recognize technology transfer activities or accomplishments as one factor on which promotion decisions for employees may depend.

Another disappointing factor was the slow pace of increase in the number of patents issued. In 1986, 314 were issued; in 1989, there was only an increase of 3—a total of 317 patents issued. The granting of exclusive licenses was no more promising. In 8 cases, the number of exclusive licenses granted in 1989 was less than the number granted in 1986. In 11 cases, there were more, and in 136 cases, there was no difference in the number of exclusive licenses in 1986 and 1989. Even though somewhat more encouraging results have come from an update to this report, there is still room for significant improvement.

Returning to comments from industry leaders, there was mention of concerns about fairness in access to CRADA opportunities. Questions of fairness and liability raise red flags at agency headquarters, and some are establishing patent policy boards to deal with these questions. If a technology may be particularly attractive, it may be safest to advertise and seek collaborators competitively, even though a CRADA is not a procurement activity and thus does not require this. The National Institutes of Health (NIH) goes a step further on the question of fairness, making recommendations on what it considers fair pricing of pharmaceuticals it transfers that are of critical impact to society.

Other industry leaders interviewed pointed out that one of the most essential improvements a federal laboratory could make would be to make it easy to contact designated officers or attorneys associated with the federal facility. Additionally, they felt federal laboratories should encourage face-to-face meetings or briefings to introduce the technology in detail, identify how far the federal laboratory is willing to take development on its own, and create funding for joint subsequent development work before transfer.

They felt that the most important factor in gaining access to technical information and personnel in the federal laboratories was regular one-on-one visits with the technical professionals responsible for the work. The quality and quantity of detailed information depend strongly on the character of the relationships developed with the federal technologists. News announcements, license offers, and technical reports are useful chiefly for establishing some degree of interest and for identifying the personnel involved. But the one-on-one contact is the primary ingredient for success.

Some business leaders who had difficulties found that the person technically able to provide information would many times not be accessible. Therefore, they suggest that the laboratory supply a list of sources by topic, including the name of the person able to provide information, along with telephone number. Realistically, however, some screening process is needed to prevent overload, making it doubtful if the publishing of individual scientists' phone numbers will be encouraged.

What is sometimes of most interest to industry is direct hands-on assistance in the use and selection of appropriate new technology, yet this is not a normal function in the mission of the labs. One exception is the manufacturing technology centers of the National Institute of Standards and Technology (NIST).

The substantial differences in institutional perspective between persons in government and industry require both patience and effort to overcome. The relationship is most effective when it is one of equals, with government scientists learning from their counterparts in industrial laboratories and vice versa. Both groups attest to the value of informal networks and personal communication. The value of electronic or paper-based information systems was acknowledged, but most people recognized that significant transfer activity requires sustained individual attention—often with "champions" working on both sides. Most industrial managers are unconvinced that technology transfer has a high priority in government and are still looking for evidence of its greater visibility within the bureaucracy and increased resources devoted to the function.

The interviews with these industrial leaders did, though, reveal several trends that bode well for increased opportunities in the area of cooperative efforts between federal and industrial laboratories. First, there has been a marked trend in U.S. industry away from a traditional reluctance to borrow ideas from others, particularly from outside the corporation. Even large, high technology firms are no longer able to meet their own technology requirements solely through in-house efforts. Second, there is a much greater awareness that technological capability available in federal programs has only been partially tapped, and that the great potential for its use in commercial applications must be given serious consideration.

CHAPTER 8

SUCCESSFUL TECHNOLOGY TRANSFER: LESSONS LEARNED

Several years ago a manager at DuPont noticed the favorable changes in federal legislation regarding technology transfer and brought it to his bosses' attention. He was given approval to look into it. That one decision has proven so successful that this person is now the technical manager for technology acquisition and has his own staff that handles tech transfer from government labs. Other large companies are beginning to follow suit. What have they learned?

Several broad lessons can be drawn and apply both to government and to industry as they seek to move federally sponsored technology into the commercial arena. For example, the government's lessons would be to (1) institutionalize the process, (2) concentrate on networks and personal contact, (3) focus on the organization's technological strength as the basis for commencing transfer activities, (4) be open to third parties, (5) seek cooperation early in the technological cycle, and (6) sustain a continuity of effort. Industry's lessons include: (1) patience, (2) persistence, (3) be discriminating, (4) seek allies, and (5) be willing to learn.

LESSONS LEARNED BY GOVERNMENT

One of the first principles is the clear need to institutionalize technology transfer. Legislation over the past several years along with positive action on the part of agency leadership support this direction. However, no amount of regulation and legislation can substitute for weaving this process into the general fabric of the agencies involved—to the extent that this function becomes a regular part of daily activities rather than secondary, add-on activity. This means that technology transfer will require at least some formal organizational structure with both visibility and influence within the agency. However, the nature of technology transfer thrives on a certain degree of serendipity and, therefore, should not be overly structured or bureaucratized.

The selection of the former head of Exxon's R & D organization as the new director of Argonne National Laboratory clearly was made to strengthen the internal culture for technology transfer. Perhaps a new type of revolving door is coming into being.

Second, those experienced in technology transfer recognize the importance of networks of personal contacts as a primary means for becoming aware of the existence of new technology. This is another reason the process cannot be as neatly organized, as is the temptation in most bureaucratic organizations.

Third, technology is more likely to be transferred where efforts are focused on areas of technological strength within the organization. Patenting is critical, and the assistance of advisory boards with members from outside the government is gaining acceptance. Technology transfer is not one-way but is truly a reciprocal relationship in which each participating organization receives technological benefits.

Fourth, because of the substantial differences in approach and perspective between public agencies and industrial organizations, opportunities for cooperative effort may be more easily developed in conjunction with a third party. This may be a professional organization, a university where neutral ground exists, or through an agent or broker. In such a case, each participant contributes to a joint effort.

Fifth, because there is less familiarity with such considerations as capitalization, depreciation, and return on investment in the government, focus should be on information exchange at the earliest stages of technology development. That is, it should concentrate on these stages *before* product or process is defined in great detail. Technical expertise is more easily applied to a spinoff problem or solution if it has not yet developed to the hardware stage. This tends more often to involve peer communication—which is a more natural channel of information exchange and cooperation.

Finally, a public agency must have both political and social support to sustain any significant, organized technology transfer activity. Continuity of effort is essential if this process is to be more than a passing fad.

Lessons Learned by Industry

Patience is, perhaps, one of the most difficult lessons for people in industry interested in technology transfer to learn or to accept. Because companies will be dealing with a public agency that often has a keen sense of accountability to its political superiors, to the legislative body, and to the public, you must be prepared for a slower pace in decisions than one might find in private industry. This is not always the case: agencies whose leadership is determined to press technology transfer have found ways to speed the decision process. Sometimes bureaucratic or legal tangles may so slow the process that the effort is no longer worth the prize. As government agencies become more comfortable and proficient with cooperative endeavors, this will prove less and less true. At any rate, management on the industry side should be willing to persevere during initial forays into technology transfer.

Perhaps second to patience is persistence. In most instances, the scientists representing both parties will find it easier to reach agreement than is true of the organizational leadership or the legal experts. Like the biblical story of the widow seeking a fair settlement from the judge, one must continue to work the system.

Industry managers should be discriminating—that is, they should be selective in the technical area where they seek assistance. This means either obtaining help or learning enough about where the relevant centers are so that they can effectively screen the many technical possibilities. As noted in subsequent chapters, quite a number of available tools are to assist in this process.

Seek allies and be willing to share technology. This works both ways: scientists from both institutions must feel they have made a contribution and are thereby further ahead technologically at the end of the process than they were at the beginning. It behooves industry to develop an appreciation of the institutional perspective of government scientists (as well as vice versa). Match peers to peers, and, as they discover mutual interest, they will become allies in the joint effort for successful transfer.

Finally, industry managers must be willing to learn. For too long there has been a nearly impenetrable wall to cooperative relationships between government agencies and business. Both sets of institutions must now learn how successful cooperation can be maximized for mutual benefit. One way to accelerate this process is to enlist the help of an experienced broker.

PART II

WHERE CAN YOU LOOK FOR HELP?

CHAPTER 9

BROKERS

You have seen the problems; you are aware of the positive changes in the government; but, you just do not have the time or the wherewithall to do the initial legwork involved in federal technology transfer. Do not give up now! There is help. Throughout government agencies, both federal and state, there are many people whose responsibility is to help you locate technology or expertise. The additional boon is that more and more of these people are gaining an appreciation of the needs of *both* sides.

THE BROKER FUNCTION—EXAMPLES OF HOW GOVERNMENT IS HANDLING IT

Although *broker*, as used in this chapter, refers primarily to a not-for-profit *organization*, these organizations are made up of *individuals*, some of whom sincerely get personal satisfaction from going beyond a normal bureaucratic response in their efforts to make technology transfer a success. The broker's job is to facilitate this transfer of technology. As such, brokers have to look in both directions—both inside and outside the federal framework. When the job calls for locating an internally developed technology or expertise, the broker uses what is termed as *inreach* or technology *pull*. A broker whose primary responsibility is inreach is most interested in outside parties' desires and preferred mechanisms of information access—assistance in defining the real technical problem or need, arranging of individual laboratory visits, cooperative research and development agreements (CRADAs), workshops, seminars, contract research or consulting, use of laboratory facilities, licensing, employee exchanges, and sponsored research. Inreach is also used internally to discover commercially valuable technology that might otherwise be missed.

Outreach—or technology *push*—involves seeking an outside party to commercialize an existing technology or to collaborate to solve a joint problem. This can involve making an innovative suggestion as to a new, or unexpected, use for a technology, and in making its application relevant and salable to a targeted commercializer. Many brokers concentrate on one area or another, but both aspects require follow-up and feedback to both parties to keep things moving and avoid intellectual property misunderstandings.

As previously pointed out, recent legislation established laboratory-level offices, Offices of Research and Technology Applications (ORTAs), specifically to assist in the location of internal technology. The brokers within these centralized information offices are aware of the original sources of valuable innovative technology and of how to access this information and its creators.

For instance, NASA's Technology Utilization Offices (TUOs) have access to the Technology Utilization Network System (TUNS)—a NASA-wide, up-to-date record of all innovations by NASA employees or those of NASA contractors. ORTAs have similar information at other agencies. Both ORTAs and TUOs are usually personally acquainted with the innovative researchers at their respective labs and know their preferences on how to be approached.

The navy's "technology ferret" program, set up in 1988, uses an industry retiree who works with the patent counsel at the Naval Underwater Systems Center to identify technologies that have commercial value. The program was modeled on a successful British program. This is another inreach approach.

National Institute of Standards and Technology (NIST) is in the process of creating a series of regional hands-on demonstration centers that will be affiliated with nonprofit institutions or organizations, which could be termed "brokering locations," for the transfer of manufacturing technology. This is an example of outreach.

As described in Chapter 3, the Federal Laboratory Consortium for Technology Transfer (FLC) was formally chartered by the federal Technology Transfer Act of 1986. Currently there are six FLC geographical regions, each with a regional coordinator. These regional FLC representatives provide an entry point for accessing the member labs, regardless of what agency they are under. This networking mechanism effectively provides rapid communications for interlaboratory cooperation.

The FLC is required to use 5 percent of its funding for technology transfer demonstration projects with nonprofit entities. Recently, the FLC announced completion of a model demonstration project that provides federal laboratories with access to the names, addresses and phone numbers of technology acquisition managers of more than one thousand companies through use of the Technology Targeting Database created by the University of Utah. A new demonstration project with the U.S. Conference of Mayors is currently under consideration.

Some brokers are allowed to charge fees for their services, like NASA's ten former Industrial Applications Centers (IACs), which have now been reduced in number to support six Regional Technology Transfer Centers (RTTCs), in the geographic regions as defined by the FLC. However, the assistance rendered by such an entity can be invaluable. For example, one specialty engineering company was directed by an IAC to a NASA expert in bolting technology, assistance that was well worth the fee. Rather than encounter resistance, a company can take advantage of this form of assistance—and get a simple answer to a technical inquiry.

The ten former IACs, mostly based in universities, serve 10,000 industries and other clients per year. NASA's software broker, the Computer Software Management and Information Center (COSMIC, University of Georgia), houses 1,500 software programs it transfers to industry for fees. IAC affiliates currently include thirty state business assistance centers, with twelve more being negotiated. In addition, IACs are developing agreements with the Small Business Administration for further cooperative efforts.

Currently, two NASA contractors—RTI (Research Triangle Institute, North Carolina) and Rural Enterprises, Durant, Oklahoma—work with industry to define so-called problem statements that might be solved with NASA know-how. Their efforts result in the creation of joint efforts with industry called the Applications Engineering Program. See Chapter 17.

Another brokering mechanism is the use of advisory boards. The Department of Energy's National Renewable Energy Laboratory (NREL) advisory board offers independent advice and counsel on NREL technologies and their commercial potential. The board, composed of up to sixteen members selected from industry, institutions, and academia, provides insight into national trends in the marketplace and offers access to those familiar with market concerns to advise federal employees on the best way to move their inventions into businesses. By such leveraging of its limited resources, NREL has found its advisory board to be an efficient way to take advantage of new channels for disseminating information on its research, for making contact with potential users, and for discovering other innovative paths to technology transfer.

NREL is operated for the Department of Energy (DOE) by the Midwest Research Institute (MRI) and offers its technology for licensing through MRI

Ventures (MRIV), a for-profit subsidiary of MRI that creates private sector opportunities for technologies developed by MRI and also provides patent search and application assistance, preparation of licensing packages, market analysis to locate licensees, and negotiation of licenses.

NREL also uses the FLC's access to private sector technology brokers. One example is NREL's relationship with Technology Transfer Conferences, Inc., a private technology exposition organizer, which was facilitated by the FLC.

The NREL ORTA works closely with the family of trade associations representing industry sectors that are potential users of NREL-developed technology. These include the Solar Energy Industries Association (SEIA), the American Wind Energy Association (AWEA), the Solar Rating and Certification Corporation (SRCC), the Renewable Energy Institute (REI), and the Passive Solar Industries Council (PSIC). Assistance in technology transfer through subcontracting relationships has been obtained from SEIA, AWEA, and PSIC.

SDIO (Strategic Defense Initiative Organization) has also used a network of advisory panels with representatives from industry, professional associations, private research institutes, and universities. Regular review meetings are held to evaluate SDIO technology and its commercial potential.

Cooperation with State / University Programs

Department of Commerce

The Department of Commerce recently awarded more than $910 million to nine state governments for programs designed to encourage technology transfer between federal laboratories and local businesses. The states involved were Arkansas, Georgia, Maryland, Massachusetts, Michigan, Minnesota, New York, Pennsylvania, and Tennessee.

The Department of Commerce has established the Clearinghouse for State and Local Initiatives on Productivity, Technology, and Innovation to help policy makers and industry obtain information on approximately seven hundred state and local technology development or technology transfer programs. Information is also provided in such categories as business capital sources, incubator programs, university research centers, technical and management assistance, and training. The goal is to allow greater sharing of programmatic details for duplicating successful efforts and avoiding mistakes that have been made elsewhere.

Federal Laboratory Consortium

The authorizing law for the Federal Laboratory Consortium for Technology Transfer specifically states that the FLC should work with state and local governments and universities. This type of cooperation maximizes the effectiveness of the limited staffs of the respective organizations and provides a familiar entry point for local businesses. Examples of state service organizations the FLC has worked with include the following:

- *Oklahoma Vocational-Technical Department.* Teachers from across the state have attended workshops describing the capabilities of the federal labs and offering training in delivering services to small companies they contact. As a measure of the success of this program, Oklahoma is now second only to California in the number of calls to the FLC clearinghouse.

- *Michigan Technological University (MTU).* The university uses community college instructors to contact small businesses in the state, informing them of the technical expertise available from Argonne and Oak Ridge National Laboratories. The University-Laboratory-Industry Brokering System also makes businesses aware of the technical services of MTU and other Michigan universities.

- *Mississippi Technology Transfer.* This program uses local community leaders and county agents with assistance from technical institutes and the TUO or ORTA at the Stennis Space Center.

- *Small Business Development Centers (SBDCs) in Washington State and Iowa.* These SBDCs offer seminars and contact points for local small businesses to gain access to the federal laboratories.

National Aeronautics and Space Administration

Two NASA centers, Goddard Space Flight Center, Greenbelt, Maryland, and Langley Research Center, Hampton, Virginia, have signed a five-year technology transfer agreement with Maryland, marking a significant step in federal-state partnerships. Each NASA center named a representative for technology transfer activities who will work with Maryland's Office of Technology Development. Goddard also has an agreement with Morgan State University, Maryland, that provides for reviews of R & D reports to match areas of technical interests. The Stennis Space Center has joined with Marshall Space Flight Center to establish close ties with the economic development offices of six southern states.

NASA is also pushing for the involvement of its university-based Centers for the Commercial Development of Space (CCDSs) as marketing arms for its Office of Commercial Programs.

Department of Defense

The U.S. Air Force is working in cooperation with the Ohio Technology Transfer Organization (OTTO). OTTO agents receive requests for technical assistance and turn them over to the Wright-Patterson ORTA if they may relate to air force technology.

In a case involving a manufacturer of titanium blades for aircraft engines, the manufacturer first contacted the OTTO agent at Lima Technical College, who then called the ORTA at Wright-Patterson AFB. Personnel in their materials laboratory helped the manufacturer identify changes that could be applied to their manufacturing process. The result was a 75 percent reduction in off-specification parts.

DOD laboratories in Massachusetts have signed an agreement with the governor to cooperate on matters of technology transfer and facilitate the CRADA process by providing briefings for local businesses on the technical strengths of the labs and the mechanisms for cooperative R & D.

National Environmental Technology Applications Corporation (NETAC)

The mission of NETAC, at the University of Pittsburgh, is to help commercialize federal, state, and private research in environmental technology. Their brokering efforts have a full-service scope, including marketing, finance, patent and licensing activities, testing, and evaluation. They also can assist in obtaining financial assistance for research and development projects.

Department of Energy

DOE has an innovative approach that is spreading to other agencies with contractor-operated laboratories—the development of adjunct organizations to handle or assist in technology transfer. A DOE mechanism provides seed money to initiate adjunct organizations' technology transfer activities with a payback provision from royalties that will ideally make the operations self-sufficient. As such, some laboratory contractors are establishing adjunct organizations to license and market new technology and software because such expertise has not traditionally been available in the laboratories. Several adjunct organizations associated with state universities have been established to broker the licensing of laboratory technology.

- *The Argonne National Laboratory (ANL)–University of Chicago Development Corporation (ARCH).* A nonprofit, joint partnership between the University of Chicago and ANL with assistance from the business school, ARCH has negotiated exclusive licensing agreements in 1987 and 1988 for ANL's helium-dilution refrigerator and its portable toxic gas detector. One of their recent ventures is the joint development and commercialization with Dravo Lime Company of a new process that removes up to 70 percent of nitrogen oxides from coal stack gases.

- *Oak Ridge National Laboratory.* Through the Tennessee Center for Research and Development, Oak Ridge has issued twenty-two licenses for technology and software developed at the laboratory.

- *Edge Technologies, Inc., Ames Laboratory, (DOE).* Established to provide an organized means jointly to commercialize technology from the Ames Laboratory and Iowa State University.

Some DOE labs are implementing other mechanisms for technology transfer. In New Mexico, both Los Alamos National Laboratory and Sandia National Laboratories have followed the lead of recent university practices and set up so-called incubator operations to nurture their entrepreneurs. Tennessee's Oak Ridge National Laboratory even has its own associated for-profit venture capital group. The director of technical applications at Oak Ridge stated that the lab has spun off seven companies in the last year. The Savannah River Company operated by Westinghouse has a relationship with the University of South Carolina that uses students to assess the commercial potential of inventions and assist with drafting disclosures.

SETTING THE STAGE FOR SUCCESS

Although there now are technology brokers in government agencies, it is equally important for successful transfer that the outside technology acquisition team be staffed by individuals with persistence and other necessary personal qualities. The selection of this technology transfer team is a critical task for the decision maker, so the factors necessary for success must be identified before selection. A survey of highly successful teams has shown that the factors for success consistently include having at least one person who stays with the project to the end; a good balance between end users and developers; people with a history of good interpersonal dynamics and networking skills; past shared work and social experiences; people who are self-motivated problem solvers; and a simple management structure with only the minimum necessary accountability, milestones, rewards and incentives. Brokering success depends on being able to work together well to solve recognized problems with creativity, open-mindedness, and optimism, and on not being stifled by a cumbersome bureaucracy.

The initial selection of which lab, adjunct organization, or broker to approach can be done primarily through accessing the abundant sources of free information on federal technology, discussed in Chapter 10.

CHAPTER 10

TECHNOLOGY TRANSFER BY INFORMATION DISSEMINATION

A wealth of information sources is available to the company seeking to use federal technologies, and many of these sources are *free*. Use these existing sources of information to locate and then initiate contact with the innovator of the technology in which you are interested or the relevant Office of Research and Technology Applications (ORTA) or government patent counsel.

Information on available technology is disseminated mostly at no cost to industry in virtually every available media, including print, computer networks, broadcasts, videos, and trade shows. The biggest problem is knowing where these sources are and how to use them. The first step might be to get on mailing lists of appropriate federal lab technology magazines and newsletters. Recommendations about the availability of such free publications can be obtained by contacting the lab ORTA.

Industries throughout the United States are reaping the rewards of the publicly available information on research done at NASA. For instance, runway grooving was developed at the NASA Langley Research Center for safety purposes at airports. Through NASA's information dissemination this technology became the basis of the commercial pavement grooving business.

Carlos Horvath, a computer scientist working at Burroughs Corp., derived enough information from a *NASA Tech Brief* article to develop his concept and then build a computer chip tester. Horvath never patented the technique because he felt the public domain information in *NASA Tech Briefs* should have been sufficient for anyone to produce his system. Unisys is still using the tester.

Another step is to attend trade shows. Federal agencies are attending, and presenting at, more and more trade shows every year, opening up excellent opportunities for private sector representatives to meet innovators and technology transfer champions or to find out how to contact them.

Many companies already are reaping the benefits of their trade show visits. For instance, at one NASA exhibit, a solar collector display led to the development of better equipment for creating portable ice skating rinks. A new NASA impact-resistant padding, demonstrated at another conference, was adapted by several companies, including Kees-Goebel Medical Specialists, for use in prosthesis padding and other medical appliances.

FEDERAL LABORATORY PARTICIPATION IN TRADE SHOWS

Agency laboratories throughout the government are not only attending workshops and conferences, but are also setting up some of their own. For instance, working with Technology Transfer Conferences, Inc., the Federal Laboratory Consortium (FLC) helps organize conferences between companies and federal laboratories that focus on specific areas of mutual interest. The Government Microcircuits Applications Conference (GOMAC) is sponsored annually by the Department of Defense (DOD), NASA, the Department of Energy (DOE), the Department of Commerce, the National Institute of Mental Health (NIMH), and the Department of Health and Human Services, to discover innovative new uses of microelectronic devices and techniques in government systems cooperatively.

The Department of Commerce's first "Federal Technology Transfer Series" was a workshop on biotechnology. There are plans to hold similar workshops in ceramics, environmental technology, industrial design, digital imaging, and high performance computing. Cooperating with the Industrial Research Institute (IRI), national conferences in transferring the government's manufacturing technology to U.S. industry are being held.

The Department of the Navy displayed its unclassified technology at the navy's first Domestic Technology TransFair in Kansas City, Missouri. This meeting included personal presentations of inventions by naval inventors, as well as several sessions open to discussion between the inventors and potential commercializers. More than seventy navy research, development, test, and evaluation commands participated. Navy patent attorneys and ORTAs were also made available to give advice.

The Agricultural Research Service (ARS) holds technology transfer meetings at ARS locations. Through these and related efforts, ARS scientists make more than sixty thousand contacts per year with users of technology developed by the agency. The objective of these meetings is to make industry more aware of the kinds of research ARS laboratories conduct, what may be currently available for transfer, what unique facilities are available, and what scientific expertise is available for technical assistance. Their goal is to have at least one technology transfer meeting at each sizable location.

In October 1989, the National Institutes of Health/Alcohol, Drug, and Mental Health Administration (NIH/ADAMHA) hosted the first NIH/ADAMHA-Industry Collaboration Forum. More than three hundred industry representatives attended. These forums are now conducted annually.

Other typical agency-sponsored conferences to look for are:

- *Federal Technology Transfer Series—Biotechnology Conference*. U.S. Department of Commerce with federal laboratories that specialize in biotechnology.

- *R & D Conference & Technology Transfer Workshop*. Sponsors: FLC, Connecticut Innovations, Connecticut Department of Economic Development, Science Park Development Corporation, and Yale University.

- *NASTRAN Advisory Group Meeting*. For user's of the NASA design and structural analysis software program, NASA Structural Analysis (NASTRAN).

- *Federal Computer Conference*. National Technical Information Service (NTIS) and other agencies.

- *FLC Meetings*. Regional and national opportunities to provide input to the creation of ever-better policies for federal cooperation with industry.

- *Technology 2000*. Now an annual exposition of technical R & D from NASA's labs, including contractor-developed technology.

- *Federal High Tech*. Most recently, "Federal High Tech '90: Interagency Firms." Sponsors: National Science Foundation (NSF) and DOD.

- *Conferences on Hazardous Material*. Held in Los Angeles, California. Sponsors: IRI with the FLC.

Many technology transfer meetings are held at the laboratory itself or at the site of the sponsoring company. At these meetings, technological and legal questions can be addressed. Federal employees are more open to these arrangements when there is an opportunity for mutual benefit. Sponsoring companies also are in favor of such meetings because they can talk with the person who can answer their questions. One such meeting at Argonne National Laboratory attracted sixty-six companies to a workshop on the use of lactic acid in bioconversion of food wastes and its subsequent use in forming biodegradable plastics.

GOVERNMENT R & D INFORMATION THROUGH DATA BASES AND E-MAIL SYSTEMS

Most technology information requests involve access to someone or some publicly available publication that can answer a technical question. You have to find someone who knows where to look and has the time to do so. Some offices are mandated to do this and others do if time permits, so you have to be patient. But many requests for assistance can be met by information from computer data base searches. Universities, Small Business Development Centers (SBDCs), and commercial and government entities can provide useful search services, along with government entities such as the Department of Commerce's NTIS, NASA's Regional Technology Transfer Centers (RTTCs) and its Scientific and Technical Information Facility (STIF), now known as the Center for Aerospace Information (CASI), the Department of Agriculture's library centers, DOD's information centers, and the Strategic Defense Initiative Organization (SDIO) Technology Application Information System (TAIS).

Perhaps the most impressive of all federal technology transfer networking is the FLC Clearinghouse Electronic Mail and Resource Directory. Established in July 1987, the clearinghouse serves an important role for both laboratories and industry requestors by assisting them in making initial contact. At present, at least seventy-one laboratories, seven advisors, and eleven FLC-related organizations have been linked by the electronic mail (E-mail) system. The Resource Directory is based on a key word listing of laboratory expertise that covers more than 168 lab facilities. In 1988, the clearinghouse reportedly responded to inquires from both large and small businesses, state and local governments, and universities in thirty-eight states. Of these inquiries, more than two-thirds were for technical advice and the rest were requests for general information. The clearinghouse resources also include access to the DIALOG data base on CD-ROM, the Federal Applied Technology Data Base, and the Technology Targeting Database, mentioned in Chapter 9, which identifies industry's technology acquisition interests and contact points.

You can use the FLC Clearinghouse E-mail system as a first step in locating a federal scientist or engineer for practically any technical need. If you plan to use their services, it is always easiest if you can technically narrow the scope of the information you seek. The FLC regularly offers training at national meetings in use of the system and in teleconferencing. Inquiries are received daily at the clearinghouse, from every sector and involving a broad range of subject matter.

These messages and requests are seen on personal computer terminals by many lab ORTAs from various agencies at the same time, increasing the ability of the federal laboratories to respond to a need for technology in the private sector. Examples of the types of messages on the FLC E-mail system include sample agreements, notifications by laboratories of technology seeking commercialization or upcoming tech transfer events, and requests from companies for help with a specific technical problem.

The goal of the newly established, and still controversial, National Technology Transfer Center (NTTC) is to become the one central clearinghouse for information on federal technology with commercial potential. Created by Senator Robert Byrd and located at Wheeling Jesuit College in West Virginia, the NTTC is thought by some to be taking over tasks originally planned for the Department of Commerce. Already budgeted at higher levels than the FLC, NTTC must prove that it is not a needless duplication. There are plans to have it facilitate national debate on emerging tech transfer issues.

The NTTC will provide toll-free access to electronic data bases and will have up to thirty people on the staff trained to answer inquiries and direct callers to the lab most likely to be of assistance. NASA has the management role in developing this project with plans to integrate access to nonsensitive parts of its own Technology Utilization Network System (TUNS).

OTHER CURRENTLY OPERATIONAL FEDERAL TECHNOLOGY DATA BASES AND E-MAIL SYSTEMS

- *TEKTRAN*. Covers all parts of the Department of Agriculture (USDA) with reports on new research findings.

- *Lablink*. An E-mail system between labs, businesses, and incubators in the New Jersey area.

- *LifeNet, NASA*. A user-friendly E-mail system for U.S. and international life sciences researchers and their space life sciences colleagues. Administered from the Johnson Space Center, Houston.

- *FS INFO*. The Forest Service Information Network—Forestry On-line network with a data base provides access to scientific and technical literature of interest to the Forest Service. The data base is accessible through the Forest Service Computer Network and almost every Forest Service unit has a local FS INFO Center.

- *USDA Online.* A leader in the federal government of on-line dissemination of information, the Office of Information (OI) makes some department research and considerable other information available nationally—and instantly—on-line through two computerized delivery services.

- *Research Results Data Base (RRDB).* Through the RRDB, specialists in the land grant university system have early access to USDA research findings. Research findings are summarized on a national data system before official publication. About four thousand reports of new research findings are currently entered in the data base. In the first six months of 1986, more than twelve thousand key word searches were made on the system.

- *AGRICOLA National Agriculture Library (NAL).* Through its responsibilities of information management, NAL purchases and processes published and unpublished international literature and subsequently produces an electronic, bibliographic data base, AGRICOLA, which is available through such commercial vendors as DIALOG, BRS, and Silver Platter's CD-ROM. Its effectiveness was proven when a biotechnology firm needed information on nutrition, feed formulation, and worldwide markets for cultured salmon. The firm was investigating the potential of a bioengineered yeast pigment that could be added to salmon diets to provide the farm-grown fish with the same red flesh coloration of their wild counterparts. Without a special feed additive, cultured salmon flesh has an off-white color that results in a substantially lower market price. After completing the research, the company marketed its product and estimated a sales potential between $75 million and $100 million annually.

- *NIST Technical Information Databases.* This data base, NIST/CARB Biological Macromolecule Crystallization, contains crystal data and the crystallization conditions for more than one thousand crystal forms of more than six hundred biological macromolecules. It is part of the National Institute of Standards and Technology (NIST) Standard Reference Data Program.

- *U.S. Department of Energy Database (EDB).* The world's largest collection of energy-related data with references to fuels and alternate energy sources, conservation and environmental impacts, and national security concerns.

- *Electronic Bulletin Board, Public Health Service.* A data base of downloadable tech transfer guidance, including model agreements, patent applications available for licensing, existing cooperative research and development agreements (CRADAs), and a resource directory of contacts.

- *Technology Applications Information System (TAIS).* A free SDIO on-line data base service to promote the commercialization of SDI technology containing more than one thousand unclassified, nonproprietary abstracts. TAIS is available to any U.S. corporation or citizen who has completed a DD Form 2345 Militarily Critical Technical Data Agreement, a form distributed and used by the Defense Logistics Agency (DLA) to identify individuals and enterprises eligible to receive militarily critical technical data.

- *Alternative Treatment Technology Information Center (ATTIC)*. A system of computer data bases developed by the Environmental Protection Agency concerning the many technologies available for hazardous waste cleanup projects. ATTIC soon will be available to other government agencies and the private sector.

- *Office of Invention Development*. A data base of NIH scientists interested in collaboration and their areas of research interest together with industry contacts. Designed to facilitate CRADA initiation and licensing.

EXAMPLES OF PRINT AND OTHER MEDIA AND SOURCES

As more and more federal laboratories focus on technology transfer, they begin to see the potential of their own work. As such, many of them actively seek partners in the private sector for the commercialization of their technological innovations. So, where do you, the private business person, look for these laboratories? There are a number of sources. For instance, the *Commerce Business Daily* and the *Federal Register* publish information about laboratories seeking CRADA partners. In addition, the *Federal Register* publishes notices about federal patent licensing applications. Technical libraries at the labs are generally for the use of civil servants and contractors, but may be a source of assistance on a noninterference basis. Other sources include:

- *NASA Patent Abstracts Bibliography (NASA PAB)*. A semiannual NASA publication containing comprehensive abstracts and indexes of NASA-owned inventions covered by U.S. patents and applications for patent.

- *NASA Tech Briefs Magazine, TSPs, and STIF (or CASI)*. Government contracts require the reporting of all new innovations that result from federally funded R & D. NASA sends theirs to Stanford Research Institute for evaluation and, once intellectual property rights are established and the company elects to allow publication, these innovations are reported in *NASA Tech Briefs*. Nearly thirteen thousand of these new technology reports have been published in *NASA Tech Briefs*. In response to information requests, 1,540,360 Technology Support Packages (TSPs) have been sent out, primarily from STIF, recently renamed CASI.

- *NIST Special Publication 763*. Describes a sampling of almost two hundred NIST research projects in which NIST would welcome collaboration with outside parties.

- *CRISP Intramural Research Index*. An annual directory published by NIH/ADAMHA of all intramural projects, cross-referenced by key word and by principal investigator.

- *Department of Veterans Affairs (VA)*. The VA has solicited via *Commerce Business Daily* to identify technology transfer facilitators. The VA also has a HyperCard data base of all projects and their results, which includes pictures, names of investigators, project purpose and status, and patent status.

- *Technology for U.S. Industry*. Released by DOE, this compilation contains a fact sheet on each of the 1987 IR–100 award-winning technologies presented annually by *R and D Magazine*, including a description of the technology and its advantages. The status of relevant patents is given, including any licensing data. Possible spinoff applications and contacts for additional information are listed.

- *High-Tech Update*. "The High-Tech Superconductivity Information Exchange Newsletter," published by DOE Ames Laboratory, Iowa State University. Additional support is provided by DOD and NSF. *High-Tech Update* also has a data base counterpart.

- *Domestic Technology Transfer Fact Sheet, U.S. Navy*. This monthly publication, which goes to a mailing list of 8,000, describes recent technical innovations and explores their commercial application potential.

- The Unified Technologies Center (UTC) at Cuyahoga, Ohio, Community College broadcasts "People and Technology" over the local NBC business radio network affiliate. A recent program featured a newly developed lubrication technology that originated in NASA's Lewis Research Center. Lewis's William Waters was interviewed regarding applications of the wear-resisting material.

- Videotapes are available from many labs. One example is on basic borate technology from the USDA and the U.S. Forest Service.

OTHER DIRECT RESPONSE SERVICES

- *National Renewable Energy Laboratory (NREL) Technical Assistance Service (TAS)*. The TAS attempts to provide direct technical assistance to requests received by phone, letter, or personal visits. In 1987 alone there were 3,063 such requests, with NREL staff able to answer 85 percent of them.

- *Sharing Product Designs*. This mechanism is for products developed by USA-CERL (U.S. Army Construction Engineering Research Laboratory) but that cannot be patented. CERL will provide designs of its nonpatentable R & D developments to firms with the technical expertise and interest in furnishing the product to the army. As an example, CERL has provided private companies with the designs for the CERL-developed PORTAWASHER at no cost, with the firm bearing all production and marketing costs.

- *National Technical Information Service (NTIS).* R & D is documented in the form of technical reports collected from all agencies, announced, and released for public purchase through NTIS, Department of Commerce. For example, in 1988, more than 16,200 technical reports were provided by the DOD alone to NTIS. The DOD is developing a data base on areas of expertise within DOD laboratories that will allow NTIS to refer inquiries from the domestic sector to appropriate experts in DOD laboratories.

Technology Information from Abroad

The USDA Office of International Cooperation and Development cites several recent accomplishments of transferring technology *into* the United States. Among them was a technique, transferred from Israel, for control of soil-borne pests by solar heating of soils. This technique was refined and used on U.S. pistachio and vegetable crops. The estimated value of this technology transfer is about $31 million.

The executive order following the Technology Transfer Act of 1986 emphasized strengthening the government's efforts to collect and disseminate such valuable information.

Private Sources of Information on Government R & D and Technology Transfer

The private sector is becoming involved in providing information on government research results. *NASA Tech Briefs* magazine, now published by a commercial venture, is one example. NTIS is seeking a joint venture with a private company to market information from the government's Computer Aided Logistics (CALS) program.

Other private sector sources include technology transfer newsletters, federal laboratory data bases and hard copy directories, interviews on radio and television, DIALOG and other accessible by for-profit computer E-mail systems, publications in professional journals and their mention in directories such as the *Science Citation Index.*

NASA Examples of Transfer by Information Dissemination

By the nature of their training and work, NASA scientists frequently write articles for professional journals, especially *NASA Tech Briefs* magazine, the in-house NASA journal. Typically, interested parties contact the NASA scientist directly for more information.

The Apollo lunar suit worn by a dozen moonwalking astronauts was a complex space system that included more than a score of features intended to assure lunar mobility with safety and comfort. One little-known feature was the use of a special three-dimensional spacer material in the boots for cushioning and ventilation. That material is being used, in modified form, as the key element of a new family of athletic shoes designed for improved shock absorption and energy return and for reduced foot fatigue.

Manufactured by KangaROOS USA, Inc., St. Louis, Missouri, the new line of shoes resulted from a two-year R & D program. The company sought to reduce athletic impact forces, which are transferred by the musculoskeletal system through the foot and lower leg, and at the same time provide lateral stability. The problem was that the two functions are inversely related.

The KangaROOS development effort was based on information from a study performed by NASA's Aerospace Research Applications Center, Indianapolis, Indiana. From this effort emerged the "Dynacoil" athletic shoe cushioning system, featuring a departure from conventional design that not only reduces impact shock and provides the requisite lateral stability, but also contributes to increased athletic efficiency. The mechanical core of the cushioning system is the "Tri-Lock" three-dimensional space fabric: an advancement of the original lunar boot material. This design produces a cushioning system that loses virtually none of its shock-absorbing capabilities throughout the life of the shoe.

One further example was the use of data from the NASA-developed anthropometric handbook by Corning Glass Works to design and develop a new line of sunglasses that proved to be extraordinarily successful in the upscale market.

Using the abundant mechanisms to narrow your search may locate a pre-existing solution for your product development. Chapter 11 examines how the transfer of already-developed technology may be accomplished.

PART III

OBTAINING EXISTING TECHNOLOGY

CHAPTER 11

PATENT LICENSING

More than half of the $110 billion spent each year on R & D in the United States comes directly from the federal government. But only 20 percent of the 120,000 patent applications filed each year with the U.S. Patent and Trademark Office come from federally sponsored research. This strongly suggests that the U.S. government is falling short of the real benefits it could reap from the investment it makes in R & D. However, many agencies are working hard to change this ratio.

This increased agency interest in the importance of patenting has created its own momentum. By law, federal agencies must share at least 15 percent of royalties with federal employee inventors. Most agencies give an even larger percentage to their employees. As such, federal employees are reaping benefits along with private industry, so they, too, are gaining an increasing interest in the commercial value of their inventions.

In the first year after passage of the federal Technology Transfer Act of 1986, NASA issued patent royalty income to 35 present and former employee-inventors in the amount of $56,944.80. In 1988, the Agricultural Research Service (ARS) was awarded 28 patents. Eighteen patent licenses were issued to industry firms, and inventions reported totalled 139, an increase of 83 percent over the previous year. The National Institutes of Health (NIH) likes to boast that they "don't patent junk" and therefore have four times the licensing rate of private patents. The list of successes goes on.

The Public Health Service (PHS) reported 90 applications filed in fiscal year 1987, about 150 in fiscal year 1988, and over 200 patent applications in fiscal year 1989. In 1988, PHS total royalty income from patent licensing was $5,164,019, of which $3,399,322 was paid to the French-American AIDS Foundation in accordance with an international agreement signed in 1987. National Cancer Institute patents accounted for most of the remaining $1,988,893.

The Department of Energy (DOE) and its government-owned contractor-operated (GOCO) laboratories issued over one hundred licenses for DOE-developed technology in 1987 and 1988. In 1987, at least $297,000 in royalties was received.

The military also is becoming more involved. In 1987, Department of Defense (DOD) laboratories received 602 patent awards and submitted an additional 504 patent applications. As of 1989, the U.S. Air Force had realized about $75,000 from royalty income. The U.S. Navy has 17 royalty-bearing licenses currently in force that paid about $13,000 in 1988, of which almost half went directly to the inventors. In 1989 royalties were expected to exceed $100,000. The chief of naval research has called on all navy R & D laboratories to market actively the products of their employee efforts. The U.S. Army Chemical Research, Development, and Engineering Center has recently filed a patent on a new anesthetic compound that does not produce the normal cardiovascular side effects.

The ARS has been issued a patent on a new process for making fabric adapt to temperature changes so that it is cool in winter and warm in summer. Patents have also been filed in seventeen foreign countries.

In a recent General Accounting Office (GAO) report, increases in licensing activity were substantiated by the 164 annual licenses in the period from 1987 to 1990 compared to only 130 licenses annually from 1981 to 1986, and in the fact that 41 percent in 1990 were exclusive compared to only 6 percent in 1981.

PATENT AND LICENSING PROCESS

Because the same basic regulations apply throughout all federal laboratories, we can use NASA guidelines to gain a general overview of the patent and licensing process for federal employee inventions. Clear authority has been provided for all federal agencies to grant exclusive, or partially exclusive, royalty-bearing licenses. This exclusivity was necessary to attract risk capital to further develop and market licensed inventions initially developed in the federal laboratories. When an invention is so made and reported by an agency employee, it is evaluated by a two-step process for its significance and patentability.

First, the programmatic significance of the invention to the agency is evaluated. To measure the significance, three factors are considered: the technological contribution the invention represents, the extent the invention may be used by the agency in its programs, and the commercial potential of the invention. If the evaluation is high in any of the three areas—with commercial potential being the most important and the determining factor if the other two areas are doubtful—a decision is made to assess the patentability of the invention from a legal standpoint. This decision is made by patent counsel of the field center or laboratory where the invention was made, who, in order to remain objective, seeks input from varying technical sources other than the inventor.

Whenever an agency files for a patent, it is made available for licensing. The agency does not prejudge the need for exclusivity. Rather, when interest in the invention is expressed by one or more responsible applicants, a determination is then made as to whether exclusive, partially exclusive, or nonexclusive licenses are needed to commercialize that invention. Each applicant is required to provide plans and intentions as to how they will achieve commercialization of the invention. The agency must publish any intent to grant other than a nonexclusive license in the *Federal Register*. (Both of these requirements are set forth in 35 U.S.C. 209.) Because the granting of a license or extension of exclusivity is well publicized in the *Federal Register*, opponents and proponents are afforded a chance to be heard, and their comments are always carefully considered—and often published.

When making the decision whether to grant exclusive or nonexclusive licensing, the government considers which avenue will benefit the public more. One important concern is which course will further R & D related to the invention. For exclusive licensing, small businesses are given preference, but the licensee must engage in substantial production within the United States.

Under the old institutional patent agreements and P.L. 96-517, the period of exclusivity to a large company was limited to five years after first commercialization. This restriction has been eliminated, and now licenses can be exclusive for the life of the patent.

Most agencies have internal publications in which they announce the availability of any invention for licensing. For instance, the most frequent method NASA uses is publication of an abstract in *NASA Tech Briefs*. However, NASA also publishes a complete catalog of all of its patents that may be available for licensing in the *NASA Patent Abstracts Bibliography*. In addition, businesses may make direct inquiries to the various NASA and other agency field centers for technology that may be commercialized.

Whenever an application for a license is received, it is initially processed by the patent counsel at the field center or laboratory that has filed for, and/or received the patent. Applications for licenses must be accompanied by a commercialization plan. The counsel will review the plans and intentions, make recommendations as to terms and conditions (including royalties), and possibly begin negotiations with the prospective licensee. Along with this initial processing, the counsel will make a preliminary recommendation and forward both to the agency headquarters for a final determination which will, in turn, publish any required notices in the *Federal Register*. If the decision is made to proceed with the license, the final license agreement is negotiated by the agency headquarters in coordination with the field center patent counsel.

All agency patent licenses require royalty payments or other consideration from the licensee that reimburses the government for its patenting costs, followed with a running royalty. There are no government standards for royalty rates. As an example though, NIH rates for exclusive licenses generally will not exceed 5 percent to 8 percent. Many agencies place greater emphasis on rewarding innovative employees than in offsetting program costs, so although royalties are distributed to the field center where the invention was made to support further licensing and technology transfer activities, a fixed percentage goes to the employee inventor. The licensee is required to file periodic reports on the progress of commercialization, (which are not subject to Freedom of Information Act requests).

As further encouragement, most agencies actively seek the participation of their inventors in the licensing of technically significant inventions. They also use an inventor as a technical and business resource during license negotiations. Inventors, because of their more complete knowledge of the inventions, are very useful in the process of seeking potential licensees, evaluating their capabilities to commercialize the invention, and structuring the licenses. In addition, some agencies offer technical assistance agreements that make inventors and other technical personnel available to assist a licensee in understanding how to make and use a licensed invention.

As a general rule, if the invention bears a direct relation to, or was made in the consequence of, the official duties of the employee, and the agency is interested in obtaining patent protection, the agency has the right to acquire title to the invention. If the agency is not interested in patenting or commercializing the invention, title may be left with the employee, subject to a license for the agency's purposes. If the invention bears no relationship at all to the employee's official duties, the agency acquires no rights at all.

The federal government always retains a nonexclusive, irrevocable, paid-up license to practice the invention or to have the invention practiced throughout the world by or on behalf of the U.S. government. This is the case for all inventions made under federally funded R & D, regardless of the agency. Because the government can "have made," it can contract with a company other than the licensee to make the invention for government use.

Non-Civil Servant Inventions Resulting from Federally Sponsored Research

All new technology, whether patentable or not, developed under government funding, must be reported to the agency sponsoring the work. Title for any new small business or nonprofit (including university) invention created under government sponsorship is vested in the small business or nonprofit. It must notify the government of its election to retain title (it has two years after disclosure to do so) and execute a license to the government for government use.

If the employee wishes to obtain the title from the government, the small business or nonprofit, by electing to retain the title itself, can then issue an exclusive license to the inventor. Such transfer of title does need approval of the sponsoring agency. Royalties earned on government-sponsored university-created inventions must be shared with the inventor with the balance going to further research or education.

In the case of inventions made by large businesses, the government retains title unless the company submits a plan for commercialization and requests that the government waive title rights. The large company has eight months after disclosure of the invention to request such waiver. Indeed, some companies do this at the beginning of contracting and are granted a waiver in advance of any inventions actually being made.

March-In Clauses

If an invention is not being fully used or commercialized in accordance with the agreement, the government can impose a compulsory license or, better, exercise its *march-in* clause under the agreement. A march-in could be triggered by a complaint by a competitor that the licensee is not meeting government obligations, or by the overriding concerns of public health or safety. The authority exists to force a cessation of a licensing agreement or to require sublicensing to a business with the capabilities of reaching the marketplace. Although the thought of march-ins can be quite disturbing to companies, this right is rarely exercised by any agency.

Examples of Licensing Success

For many years, glass was the most commonly used material in eyeglass lenses. Its principal advantage was that it resisted scratching; its main disadvantage was that it was brittle and breakable, making it dangerous for the wearer. With this in mind, in 1972, the Food and Drug Administration (FDA) issued a regulation that all sunglasses and prescription lenses must be shatter-resistant, causing the use of plastic lenses to increase dramatically. But the new plastic lenses carried their own set of problems—even with delicate handling many types of plastic lenses develop visibility-reducing scratches.

Foster Grant Corporation, of Leominster, Massachusetts, spent more than a decade of research effort looking for a coating that would eliminate the scratching problem while retaining the safety advantages of plastic lenses. They finally found the answer through a cooperative effort with NASA, combining the technology of industry and government. NASA's part of the solution was a highly abrasion-resistant coating developed by Ames Research Center for plastic surfaces of aerospace equipment. The result is the Foster Grant SPACE TECH Lens, manufactured under license from NASA, now part of a multimillion dollar business. The SPACE TECH Lens surpasses glass in scratch-resistant properties and has five times better scratch resistance than the most popular corrective lenses.

New NASA patents related to an improved process for rendering glass scratch resistant by providing diamondlike coatings have been the subject of recent exclusive licenses, which should prove to be just as lucrative to the licensee as was the Foster Grant license.

U.S. Army Licensing Activity

Well into the push of licensing activities, the U.S. Army has two separate technologies already licensed to companies in the private sector. The CERL-invented Weld Quality Monitor (WQM) and the ceramic anode (CERANODE) have been transferred by granting exclusive licenses to APS Materials of Dayton, Ohio (for CERANODE) and National Standard Corporation of Niles, Michigan (for WQM).

In another program, the U.S. Army's Harry Diamond Laboratories (HDL) has signed a CRADA with Logical Technical Services (LTS) Corporation of Trenton, New Jersey. HDL has been issued one patent and has a second patent pending on a new system of laser microscopy which this cooperative research and development agreement (CRADA) will transfer into the marketplace. Separately, HDL will also license to LTS the existing government-owned patent(s), and will allow a brokering arrangement for the entire intellectual property package.

U.S. Air Force Licensing Activities

The U.S. Air Force has signed an agreement with DayChem Corporation that includes both a CRADA and a licensing agreement. According to the CRADA, DayChem will commercialize a new polymer synthesis process developed to produce thermoplastic aromatic benzoxazole polymers. DayChem will develop practical applications for the process, while working closely with the air force inventor, who will provide up to eighty hours of technical consultation.

Don't Just Look at the Obvious Agencies

Sometimes the least obvious agencies may be the best place to look for a commercially valuable patent. For instance, the Bureau of Reclamation is an unexpected source of new technology for licensing. Through licensing agreements with this relatively obscure agency, new technologies have emerged not only in the areas of underground surveying (mining applications), but also in electric power generation and transmission control systems.

Early involvement, through CRADA, is a good way to assure up-front licensing of patentable solutions *before* they are made available to your competitors and can be a definite plus in applying for preexisting background patents. One area yet to be tried in brokering for federal labs is working with for-profit brokers in a royalty-sharing arrangement. Most probably, this will first work with contractor-owned technology, as is the case with Midwest Research Institute Ventures and DOE's National Renewable Energy Laboratory.

CHAPTER 12

SOFTWARE

Evolving Laws and Regulations

Historically, software designed by federal employees has not been subject to copyright, but this may soon be changing. Currently, technical data and computer software generated by federal employees cannot be protected as an intellectual property right other than to the extent the subject matter is patentable (and patented). Thus, it is available for unrestricted dissemination unless it is classified.

This leaves a lot of gray area. For instance, a recent court case overruled Ashton-Tate's right to sell their dBase product under a copyright because the original software was developed at NASA's Jet Propulsion Laboratory and was thus not subject to uncopyright at the time of creation. In another example, although some federal software dissemination programs request that their customers sign a licensing agreement limiting the number of copies they can make, does the law of contracts override the fact that no one holds the authority of copyright? At present, consideration is being given to amending the laws to allow computer software developed by federal employees to be copyrighted and transferred to the private sector under royalty-bearing licenses.

The government can, though, assert copyright to software developed by a contractor at the government's direction. A contractor can also copyright software it develops based on commercialization plans submitted to the contracting officer and patent counsel of the relevant laboratory or agency.

SOME CHANGES ARE BEING MADE

TRAINING TECHNOLOGY ACT

One little-known law currently in effect concerning federally developed software is the Training Technology Act. As section 6101 of the Omnibus Trade and Competitiveness Act (1989), it mandates the transfer of federally developed training technologies and resources for all federal agencies. Actual implementation of this law is questionable, yet there are signs of activity. For example, the air force has entered a five-year agreement with the Lee Iaccoca Institute at Lehigh University to transfer its intelligent tutoring technology to the public classrooms. This is built on the $15 to $20 million prior investment in these tools by the air force.

GOCO SOFTWARE COPYRIGHT POLICY

In June 1988, a new policy for government-owned, contractor-operated (GOCO) contractors to copyright and license computer software was issued. Such contractors can now convey commercial rights to prospective licensees, thereby encouraging the commercialization of government-funded software developed at contractor-operator laboratories. As with patents, the federal government retains the royalty-free right to use the copyrighted software for its purposes and to distribute printed copies of the source code.

EXAMPLES OF SOFTWARE CURRENTLY AVAILABLE AND SOURCES

- *Federal Computer Products Center (FCPC)*. Operated by the National Technical Information Service (NTIS), FCPC sells software developed at federal laboratories.

- *HAZARD 1, NIST*. This National Institute of Standards and Technology (NIST) computer program with a new fire hazard analysis method for reducing fire losses and costs is available from both the National Fire Protection Association and the NTIS in Springfield, Virginia.

- *Supertrapp, NIST*. This personal computer data base used for determining density, viscosity, and other engineering properties of hydrocarbon mixtures is available from the NIST Standard Reference Data Program.

- *ChemMap, Department of Energy (DOE)*. Software for second surface analysis. ChemMap surface analysis software, developed by Lawrence Berkeley Laboratory (LBL), successfully translates one-dimensional profiles from spectroscopic imaging detectors into two-dimensional surfaces, and does this 15,000 times faster than previous methods.

- *BLAST, DOE*. The Building Loads and System Thermodynamics (BLAST) analyzes energy use for buildings. Over fifty firms presently use it. The University of Illinois has been established as a BLAST support office to assist users.

- *Army Software for AUTOCAD and Training*. The "Teaching Assistant" program assists AUTOCAD software users in learning concepts for automated drafting systems. It is currently being marketed by Electronic Courseware Systems (ECS), Inc., of Champaign, Illinois, under a CRADA with the U.S. Army Construction Engineering Research Laboratory (USA-CERL) of Champaign, Illinois.

- *Army–ISICAD CRADA*. ISICAD, Inc., and CERL have signed a CRADA to incorporate software developed by CERL into ISICAD's existing three-dimensional CAD software, CADVANCE.

- *Soil Conservation Service (SCS) Computer Software*. SCS has four engineering and two economic programs available through NTIS (Department of Agriculture).

- *NREL Software, DOE*. Scientists at the National Renewable Energy Laboratory (NREL) have developed Building Element Vector Analysis (BEVA), a computerized method of determining a building's energy performance and peak loads.

- *Department of Veterans Affairs*. Surgery simulation and modeling software has been developed at the Rehabilitation R & D Center in Palo Alto, California, that allows the visualization of the results of surgical options before operating.

- *COSMIC (Computer Software Management and Information Center)*. Located at the University of Georgia, COSMIC sells software developed at NASA. Maintaining a vast software library, COSMIC gives industry access to thousands of computer programs developed for NASA and the Department of Defense (DOD), as well as selected programs for other government agencies. Available programs range from management to information science (retrieval systems) and computer operations (hardware and software). Hundreds of engineering programs perform such tasks as structural analysis, electronic circuit design, chemical analysis, and the design of fluid systems. Other programs determine building energy requirements and optimize mineral exploration. COSMIC services go beyond the collection and storage of software packages. Programs are checked for completeness, special announcements and an indexed software catalog are prepared, and programs are reproduced for distribution. Customers are assisted in identifying their software needs, and follow-ups are made to determine the successes and problems and to provide updates and error corrections. In some cases, users are offered guidance from NASA engineers in installing or running a program. Information about programs described in *NASA Tech Briefs* articles can be obtained by contacting COSMIC.

- *NASTRAN (NASA Structural Analysis)*. Although this program was originally developed in and for NASA, it has had literally thousands of variations and uses throughout U.S. industry, particularly in design and testing.

- *AdaNET Software Repository*. Not yet completed, this West Virginia repository will transfer Ada language software (principally to come from NASA and DOD).

- *Intelligent Physics Tutor, NASA*. Developed at the Johnson Space Center as an Applications Engineering Program (AEP), this program uses an expert system to teach problem solving in high school and entry-level college physics. Partners included Apple Computer, the University of Houston, and the Clear Creek, Texas, Independent School District. Additional funding support came from the Brown Foundation, Pennzoil Products Company, and the Texas Advanced Technology Program.

Generally, the prices charged for government software are only supposed to offset the costs of their dissemination and not their development. There are, though, wide variations and a fair amount of arbitrary price setting.

If transfer of a copyright is essential, it does not exist yet for civil-servant developed programs. Find out if a contractor actually created the program and encourage them to copyright their work and present to their sponsoring agency a plan for commercialization before it becomes public domain. If copyright is not important, a wealth of software is available in every field.

CHAPTER 13

CONTRACTOR-OWNED TECHNOLOGY

Because the majority of federal R & D dollars is spent by contractors that are not traditionally interested in technology transfer, there is a vast untapped source of new technology in federal agencies. As mentioned in Chapter 11, under current law, small companies have two years to elect to retain rights to inventions made under federal contracts. Large companies have eight months after disclosure in which to request a waiver of the government's right to title. This process also offers data security for technology a company develops under government contract, allowing some measure of protection for the company from the Freedom of Information Act, and keeping the new technology a trade secret while the company decides whether to patent.

EXAMPLES OF TRANSFER OF CONTRACTOR-OWNED TECHNOLOGY

Charles Kelman, a New York surgeon, is the developer of the KELMAST phacoemulsification technique for restoring the eye's focusing capability after cataract surgery. Kelman wanted to improve the process, but he did not have access to the appropriate technology until he read an article in *Forbes* magazine about Aura Systems, a Los Angeles company specializing in magnetic suspension technology under contract to the Strategic Defense Initiative Organization (SDIO). When Kelman contacted them, Aura Systems realized the commercial potential of the application of their technology to improving KELMAST. Kelman and Aura Systems together created Aura Medical Systems, Inc., to market the new KELMAST. They are now exploring such possible future applications as the removal of plaque from arteries and obstructions from organs, and for reaching other inaccessible parts of the body without invasive surgery.

TECHNOLOGY TRANSFER FROM GOVERNMENT-OWNED, CONTRACTOR-OPERATED LABORATORIES (GOCOs)

Significant changes are occurring in the seriousness with which the operating contractors of government labs are handling technology transfer. This is due to the waiver of government title rights granted the Department of Energy (DOE) lab operators and to some innovative contract negotiations.

Martin Marietta Energy Systems, Inc., the operator of DOE's Oak Ridge National Laboratory, has a contract with the government that pays for the initial operating expenses of technology transfer. The government will be reimbursed through royalties, eventually making the program self-sufficient. This no-risk arrangement is one of the only ways that large prime contractors will take technology transfer seriously, and it most likely will be adapted by other facilities and companies if it proves viable. Traditionally, technology transfer has not been a part of their business plan. Now that it is in place, it is making a difference to their employees in that meaningful recognition, cash awards, and even royalty sharing have been implemented. (However, do not expect the government to make such an offer to small companies.)

As an example of the results, under a DOE waiver of patent rights as part of DOE's technology transfer program, Martin Marietta Energy Systems, Inc., has granted Performance Technologies, Inc., of Lynchburg, Virginia, a license to market a device that can diagnose the condition of electric motor-driven systems while they are running. Martin Marietta claims no interest in funds received under the license. Instead, under a DOE-approved plan, all of Martin Marietta's royalties will be used to support the technology transfer program at Oak Ridge National Laboratory. Because licensing profits are not a goal, the royalty rates may be lower than industry standards.

Another technology available from Martin Marietta consists of a simple method of polymer entry that renders alkaline, cement-based grout containing liquid waste impervious to attack by even concentrated acid. There is potential for applications using other materials and various polymers. A dedicated program to market these inventions will bring them to the attention of potential licensees much sooner than might otherwise be expected, which often makes the difference in successful commercialization.

National Renewable Energy Laboratory (NREL) technology available for licensing includes a window with the insulating value of a solid wall. The patented NREL vacuum window works like a thermos bottle. A thin, evacuated space is permanently sealed between two glass panes held apart by small glass beads, creating an effective barrier to heat flow. NREL is operated by Midwest Research Institute (MRI) with its patent licensing handled by a for-profit subsidiary, MRI Ventures.

Researchers at DOE's Pacific Northwest Laboratory have developed a sensor for detecting water and organic liquid contaminants within the ground that is available for licensing. It is said to be less expensive, simpler to use, and more accurate than existing devices.

Acknowledging how vital it is to U.S. international competitiveness that new superconducting technologies be transferred quickly from national laboratories to industry, DOE's Argonne National Laboratory and American Superconductor Corporation have entered into an exclusive licensing agreement regarding a new technology for making superconducting wire. The agreement is one of the first to license superconducting technology from a government laboratory to private industry. Under a separate cooperative research and development agreement (CRADA), American Superconductor is providing $100,000 to fund Argonne research on other superconducting technologies.

Sandia National Laboratories (SNL) have signed an agreement that grants Perma Charge Corporation, of Albuquerque, New Mexico, an exclusive license for a microcellular polymer foam. The agreement allows for payment of royalties to SNL and gives Perma Charge an eighteen-month option to take a license also on use of the foam for liquid filtration.

SNL has a policy to make 0.5 percent of its budget available as technology maturation funds similar to NASA's funding of its Applications Engineering Program. See Chapter 17.

The University of California Berkeley, the Lawrence Livermore National Laboratory (LLNL) operator, has granted licenses to three companies for a new ceramic-metal material, "Cermet." It is expected to have the largest commercial market of any technology yet developed at LLNL. It is well suited for wear parts, particularly earth drilling bits, electronic components, and cutting tools. As more than two hundred other companies have indicated an interest in licensing this new material, it is expected to be the most successful technology transferred from LLNL to date.

Air Force Policy for Sensitive Technology Owned by a Contractor

The U.S. Air Force strongly recommends that a contractor first obtain air force clearance for specific technologies that can be commercialized without adversely affecting national security before commercializing or licensing to another company.

The Technology Transfer Opportunity from Small Companies

Although small business is given preferential treatment to help create commercially successful enterprises, the longer their primary source of income is the government, the more difficult it seems to be for small businesses to become tax revenue generators rather than spenders. Aggressive programs to assist in licensing efforts can help. By taking advantage of federal and state technology transfer resources, these small contractors, including Small Business Innovation Research (SBIR) grant winners, can assess their inventions and do outreach. For instance, an Office of Research and Technology Applications (ORTA) may recommend potential licensees, and lab employees may assess the technology. This early identification of licensing candidates will help small contractors decide whether to retain rights and pursue patenting (if the contractor elects to retain title, the government will not pay for patenting costs).

CHAPTER 14

SPINOFF COMPANIES AND EMPLOYEES

Spinoff companies and employees are the best source of know-how to go with the technology. But it is a commercial risk unless the innovator can work with people who have business expertise. Assistance is often needed to assure that spinoff companies get the technology to the company that can *sell it* to *its* market.

The following are some excellent examples of how the spinoff of employees to create new companies has been both successful and highly profitable for those involved.

NASA SPINOFFS

Many times, NASA employees see the potential for a particular technology they are using that goes beyond NASA. If these people have been primarily responsible for the development of that technology, NASA frequently encourages them to pursue commercialization of the technology by assigning patent or license rights to the employee(s).

One good example of this is the electrostatic precipitator control developed at Langley Research Center and licensed to Kinetic Controls. Another is the digital-imaging technique originally developed from technology at the Jet Propulsion Laboratory, and now marketed in a stand-alone chromosome analysis work station by Perceptive Systems, Inc., headed by two former NASA employees.

Dr. James Laudenslager and two other collaborators on a project left NASA's Jet Propulsion Laboratory and Bell Laboratories to form a new company, Advanced Interventional Systems, Inc. (AIS), whose mission was to develop the excimer laser and its associated fiberoptics for use in plaque removal from human blood vessels.

NREL-Spawned Spinoff Companies

Recognizing that new businesses based on technologies developed by National Renewable Energy Laboratory (NREL) are a valuable contribution to the country's economic development and an effective way to transfer research to the private sector, more than twenty new companies have emerged from the institute in the last decade.

A Shining Example of Success

As experienced entrepreneurs and investors know, for successful technology transfer, it is important to involve the inventors or developers. One way to ensure their involvement is to hire them. Accordingly, the founders of Tolfa Corporation hired, from the Department of Veterans Affairs (VA), the principal investigators of two separate projects and are working intensely toward commercialization, as only a small, enthusiastic start-up company can. Computerized Visual Communication (C-VIC), now named Lingraphica, will be the first product manufactured and marketed by Tolfa.

The project began on August 30, 1990, when the Palo Alto VA Medical Center and Tolfa Corporation, a Palo Alto, California, firm, signed a cooperative research and development agreement (CRADA) for final development and commercialization of the Desktop Vocational Assistant Robot (DeVAR). DeVAR had been under development at the Rehabilitation R&D Center for eleven years. The agreement gave Tolfa exclusive rights to manufacture and market any patentable devices developed under the agreement. In return, Tolfa pays the VA royalties on sales of the devices. These royalties will be shared with the federal inventors or co-inventors, as provided by the Technology Transfer Act of 1986.

Discussions leading to this agreement began when Dr. Humberto Gerola, now president of Tolfa Corporation, inquired about Rehabilitation R&D Center projects. Dr. Gerola had the idea of establishing a company to develop and market rehabilitation products. Over a period of several weeks, Dr. Gerola brought potential investors into the R & D center to talk with investigators about their various projects.

Two products were finally selected for commercialization, the C-VIC system for severely impaired aphasics, and the DeVAR vocational robotic work station. In the case of C-VIC, patent rights had already been released to the inventors by the government. A private licensing agreement was negotiated directly with VA coinventor Dr. Richard Steele. Because changes in its design are still under way, no determination of DeVAR patent rights has been made yet. The cooperative agreement between the VA and Tolfa stipulates a three-year collaborative effort to bring DeVAR to market.

Facilitating the Process

In an attempt to facilitate the spinoff process, agencies are giving preference to employee inventors if no other company has produced a commercialization plan for new technology. If a company does apply for a license, but employees can show the best business plan, they will still receive preference. As such, new technologies are leaving the labs with know-how, not just patent rights.

Another significant improvement would be to provide financial security for credible commercialization plans of employees willing to form new companies. This would encourage more employee inventors to step forward.

Some labs are cooperating with local business incubators actually to assist when employees want to leave the lab and start a business. An option in the beginning is to allow the federal employee to continue in the government job for a while and act as a consultant to the commercializing company.

PART IV

WORKING TOGETHER TO CREATE NEW TECHNOLOGY

CHAPTER 15

COOPERATIVE RESEARCH AND DEVELOPMENT AGREEMENTS

Cooperative research and development agreements (CRADAs) represent one of the newest mechanisms to promote the transfer of technology from federal laboratories to commercial and other uses. Essentially, the 1986 Technology Transfer Act extended to all federal agencies the capability previously held by only a few, such as NASA and the Department of Energy (DOE), to engage in cooperative research on a very close basis.

WHAT IS A CRADA?

First and foremost, a CRADA *is not* a procurement instrument. The authors of the Technology Transfer Act of 1986 were careful to make this explicit, so that a cooperative process could be initiated without the administrative red tape, time-consuming competitive bids, and related restrictions that are often a part of the procurement process. Actually a cooperative venture is not how a federal agency purchases or procures something. Instead, it is truly a way to extend federal laboratory missions through the process of technology transfer.

A CRADA absolutely requires *mutual* interests on the part of the participating laboratory and the cooperator (an industry, university, not-for-profit, or state or local government entity). Both parties are expected to bring skills, know-how, and resources to this joint effort, the only restriction being that the federal laboratory cannot pay public funds to the cooperator. However, the collaborator can use any materials, facilities, personnel, and the like in this process.

Another important aspect of this mutuality is that the effort must fall within the scope of the laboratory's primary mission. This usually does not present a problem, because if the laboratory has useful technology to transfer, the pursuit of that technology toward an application nearly always will add to the perfection of that technology and thereby increase the laboratory's capability and understanding. In most instances, the commercialization of a particular technology falls within the interest of the laboratory and the furtherance of its general mission, although the laboratory rarely is charged with the responsibility for such commercialization.

Section 12(b)(1) of the Technology Transfer Act of 1986 specifically authorizes a federal laboratory to "accept, retain, and use funds, personnel, services, and property from collaborating parties and to provide personnel, services, and property (but not funds) to collaborating parties." Such "collaborating parties" include other federal agencies, units of state or local government, industrial organizations, partnerships, and limited partnerships, and industrial development organizations, public and private foundations, nonprofit organizations (including universities), or other persons.

Thus, the door is open for a cooperative research venture of mutual interest, falling within the general mission of the laboratory, and not involving the use of federal funds being paid directly to the collaborator or collaborators.

Examples of CRADAs

Of approximately five hundred CRADAs currently in effect, at least two-thirds represent agreements entered into either by the National Institutes of Health (NIH) or the Department of Agriculture—both of which have pioneered in the use of this instrument. NIH has been preeminent in perfecting model agreements, beginning with work in 1988 when it proposed a basic five-page standard agreement covering the rights and responsibilities of the parties involved, how information was to be exchanged and used, the determination and disposition of intellectual property rights, and so on. The key attachment to such an agreement usually is an explicit statement of work defining the technical goals, activities, and responsibilities of the parties involved. The bulk of NIH CRADAs deal with pharmaceuticals, biologicals, and special chemical compounds used for testing in both laboratories and medical facilities.

One of the early CRADAs developed by the Agricultural Research Service (ARS) was with the company Embrex, Inc., for the purposes of developing an automated system for inoculating chicken eggs to prevent Mericks disease. ARS had developed the vaccine, and Embrex had developed an experimental technique that would provide low-cost and automated means for inoculating the eggs. A CRADA brought the two together to make a successful technological advance.

In quite a different field, the Bureau of Reclamation of the U.S. Department of the Interior recently entered into a CRADA with Woodward Governor Company jointly to develop a multiple input controller for hydraulic turbines—ultimately leading to a commercial product that could be used widely in hydroelectric power installations.

Recently the DOE entered into a CRADA with Life Technologies, Inc., (LTI) for the first government-industry cooperative research pact as part of the Human Genome Project. LTI will provide modified enzymes, natural chemical catalysts, and nucleotides in the process needed to sequence DNA. This should put the company in the forefront of this important effort in biotechnology.

In another example, the army's Civil Engineering Research Laboratory (CERL) entered into a CRADA with ISICAD, Inc., a software company working in computer-aided design. ISICAD worked closely with CERL engineers to learn a much more user-friendly system of computer-aided design software that the army had developed. ISICAD incorporated elements of the army program into its own technology to create a new product it is currently marketing.

Finally, the air force's Rome Air Development Center, which has become a focal point for photonics research, recently entered into a CRADA with the New York State Science and Technology Foundation. This CRADA will provide the basis for a joint endeavor to further photonics development toward commercial applications through a consortium under the New York State entity.

In each of these cases contact was made in a variety of ways but frequently involved scientist-to-scientist contact and assistance and support from the Office of Research and Technology Applications (ORTA).

How to Begin the Process and Make It Successful

The CRADA is an excellent way for a company to become closely involved in a particular technology in a federal laboratory. It has the added advantage that agencies and laboratories are authorized to make agreements before development begins regarding transfer of intellectual property rights. This means, at the very least, that the cooperator can have first choice on a possible exclusive license to practice the jointly developed technology.

An absolutely key requirement is that the potential cooperator has the technical capability to work with the particular federal laboratory. Without that, the cooperator really has very little to contribute—even when the principal objective is to commercialize the product.

As was the case in DOE's Genome Project, an agency seeking cooperators to commercialize a technology often will advertise such an opportunity in the *Commerce Business Daily*, requesting expressions of interest and providing further material for response to a specific set of requirements by the agency. The purpose is to ensure that all potential cooperators have an equal opportunity to be informed about such proposed cooperative endeavors.

A quite different approach can be made if the cooperator has a particular technological interest and is able to locate a laboratory or laboratories that can deal with that commercialization problem. In such cases where the private cooperator makes the overture to the federal laboratory, the opportunities do not need to be advertised, and it can be strictly a bilateral deal.

Experience demonstrates that the most successful CRADAs have involved scientist-to-scientist communication in significant detail, even before institutional involvement by laboratory or company management. Sooner or later management must be brought into the negotiating process, but the first criterion usually is to identify scientific or technological mutual interests.

Several other general guidelines can help ensure success. First, both parties should express their concerns at the beginning of the process so that the practicability of entering into a CRADA can be determined as early as possible.

Second, it is desirable to limit disclosure of proprietary information to only what is essential to the conduct of the mutual project. Information that is a part of the joint project can be protected from exposure (such as through the Freedom of Information Act) for periods up to five years. However, because many scientists in federal laboratories are not used to dealing with proprietary information, it is always a good practice to disclose only what is absolutely essential.

Third, one should keep the agreement simple and explicit. As NIH and other experience suggests, the standard terms can be kept to a minimum, usually five to eight pages. The technical task undertaken should be clearly described, including a detailing of what specifically the various parties are to contribute, what the goals are, how they are to be reached, by whom, and in what time sequence.

Finally, both parties must remember that this is a cooperative, mutual effort. Retaining such an environment is essential to the success of such a project, even though technical goals may not be fully reached.

CHAPTER 16

USE OF UNIQUE LABORATORY FACILITIES

There are many billions of dollars of investment in federal laboratory facilities and it would be a mistake for industry or others to neglect the potential for their use. These facilities were developed initially to meet the research needs of the various agencies or departments. Sometimes these needs included the possibility of outside users as well, particularly university or contract researchers working in areas of high priority to the agency. However, in many instances facilities are not used to the fullest extent possible, leaving the potential for nonlaboratory use from time to time.

EXAMPLES OF FEDERAL LABORATORY FACILITIES

Some of the facilities most heavily used by outsiders are at the national laboratories within the Department of Energy (DOE). From the days of the Atomic Energy Commission, a precursor to the current DOE, many costly facilities were developed with the idea that they would be open to use on a priority basis to investigators from universities, industry, and other agencies where that research was of interest to the department. Among DOE facilities specifically reserved, at least part of the time, for outside use are the High Temperature Materials Laboratory at Oak Ridge National Laboratory and the National Synchrotron Light Source at the Brookhaven National Laboratory.

Like DOE, NASA has a collection of national facilities, among which are a variety of wind tunnels that can be used not only for testing aircraft, but also for space and other vehicles as well. Recently, NASA has built several drop towers that permit limited simulation of a gravity-free environment. Several of NASA's wind tunnels are unlike any others in the world, either in private companies or other governmental facilities.

Within the Department of the Interior, the Bureau of Reclamation's primary laboratory located in Denver, Colorado, has working models of reservoir systems and dams by which to simulate and test various hydraulic and related phenomena.

The Department of Defense (DOD) has a wide variety of unique facilities that include air force wind tunnels, navy test basins, and so on. The army has a variety of testing facilities at proving grounds or similar locations that provide opportunities for field testing engineering equipment, vehicles, and other equipment with dual civil and military uses.

The National Institute of Standards and Technology (NIST) is the nation's repository for physical standards and has an extensive reference archive to include many dimensions from the tiniest to the largest involving distance, volume, and time, among others. NIST also has developed a model manufacturing facility where users can test and evaluate a wide variety of manufacturing processes under controlled conditions.

The Department of Agriculture has several unique facilities, among which is the National Plant Germ Plasm System. The only long-term seed storage facility is the National Seed Storage Laboratory at Fort Collins, Colorado, which maintains backup seed samples of germ plasm contained in working collections. In addition there are four Regional Plant Introduction Stations that keep working collections of seeds and ten National Clonal Germ Plasm Repositories having specific collections of germ plasm.

The Forest Service also has maintained a number of experimental forests, with species having been closely observed for nearly one hundred years. In addition the Forest Products Laboratory in Madison, Wisconsin, maintains an archive of woods from around the world, principally for identification purposes. Occasionally this collection is used for forensic purposes, such as the identification of the source of the wooden ladder that was used in the Lindbergh kidnapping case.

AGENCY POLICIES AND MEANS OF ACCESS

Such unique facilities are not open continuously to any requestor. The standards and reference archives typically are more easily available than are special research facilities. However, the potential user should make arrangements well in advance for visiting such facilities. Throughout the laboratory system, legitimate outside users are accommodated only if space is available at the research and testing facilities. Second, and closely related, is the principle of *noninterference*. Simply, this means that the particular schedule of a potential user must fit the schedule of the in-house or regular users of the facility, and that it will not interfere with the laboratory's activity. A third principle is that the facility sought for use is not available privately in other locations. The federal government does not compete with profit-making or revenue-producing facilities elsewhere.

A fourth guideline relates to the terms for reimbursement. Frequently researchers can use a federal laboratory facility without charge if the researcher makes available the data resulting from that use to the group responsible for maintaining and running the particular facility. In nearly all instances, the in-house researchers welcome such additions to their data base as another way to further their primary research. However, sometimes an organization wishes to maintain the proprietary nature of the laboratory tests. This will be permitted if all other conditions are met, but the user will have to pay the cost accrued by the agency in making the facility available.

Finally, some facilities are open only occasionally to outsiders, and then to a very narrow clientele that usually is closely associated with the kind of research being conducted at the facility. Examples are some of the DOE high-powered accelerators, which often are oversubscribed in terms of investigators who wish to use them, and some of the NASA wind tunnels. For instance the NASA $40' \times 80'$ wind tunnel is unmatched by any other in the world for size and speed. This facility permits the full-scale testing of some vehicles, and the demand is so great that access is almost always strictly limited to use by NASA, the U.S. Air Force, or their contractors.

Agencies vary significantly in their written policies and standard means of access. Some agencies such as DOE, NASA, and the Department of Commerce provide relatively standard approaches to the use of their facilities. In addition, some of the agencies publish guides that list the kinds of facilities and their general capabilities that may be available for use by other researchers. Like access to the laboratories themselves, the best initial point of access to facilities is to contact the Office of Research and Technology Applications (ORTA) in the particular laboratory. ORTA can give specifics on the facilities at that laboratory, and it can also probably furnish useful information about other facilities within the agency.

CHAPTER 17

OTHER MECHANISMS FOR WORKING TOGETHER

Several other means are available for gaining collaborative assistance from the federal government and its employees. Arrangements can be made for employee exchanges, consulting, jointly funded applications engineering, and for obtaining guidance on applying for technology development grants.

INDUSTRIAL GUEST INVESTIGATORS

One of the many effective ways industry can work with government in the development of new technology is through what some agencies call an Industrial Guest Investigator (IGI) program. IGI programs can work in either of two ways. For a specified length of time, company employees can work in a federal laboratory, or government employees can work in an industrial facility—all the while remaining employees of their sponsors. Most agencies are open to these arrangements, but some are more practiced and comfortable with them than others. As with any other business arrangement, a clear, written agreement is essential to the smooth flow of an IGI program.

NIST

The National Institute of Standards and Technology (NIST) has long been effective in transferring its technology through encouraging industry to assign researchers to temporary duty in NIST labs or by allowing NIST employees temporary work assignments in private facilities. NIST believes that "technology is in the minds of the people," and it shows. Over the years, as many as nine hundred industry-sponsored researchers have worked in one of the NIST labs.

NIST makes its research and facilities accessible to U.S. industry through the institute's Industrial Research Associate Program. The program is successful because it encourages individual achievement. This encouragement is evident in the published NIST program, stating that NIST and the U.S. government waive any rights to inventions made by research associates while working under the program, and acknowledges that research associates or their sponsors have the option to retain title to any of their employee inventions.

The program is paying off—both for industry and for NIST. For example, since 1982, thirty-five companies have paid more than fifty of their researchers to work in the NIST Automated Manufacturing Research Facility (AMRF). In addition, the companies provided, through loans or donations, $4.6 million worth of equipment and software to the program. The research focuses on experimenting with the methods by which robots, computers, and machine tools from different manufacturers can be synchronized in an integrated system through standardized interfaces. The other primary goal is finding a means for quality control in fully automated factory environments by improvements to the measurement process.

NREL

National Renewable Energy Laboratory (NREL) is another agency that is reaping the benefits of IGI programs. NREL each year invites more than one hundred scientists from both university and industrial organizations and from many foreign countries, as well as the United States, to work alongside NREL scientists. The number of scientists is increasing. During 1987, NREL hosted 137 visiting researchers, and in 1988 some 146 participants were at the laboratories for exchanges of three months or more.

Agricultural Research

The Northern Regional Research Center (NRRC) recently hosted three scientists who were hired by a major agricultural company and stationed at the NRRC lab.

They worked directly with NRRC staff who were assigned to three specific research projects of particular interest to the agricultural company. NRRC found the IGI program to be an excellent approach in that it accelerated the pace of the projects without burdening federal resources, and it gave the company the opportunity to enter the technology transfer process at an early point in its commitment.

NASA

In an interesting variation, the Langley Research Center is hosting a University of Oklahoma faculty member who is being sponsored by General Dynamics Corporation for two summers.

Consulting

Many federal employees work as consultants for industry firms as a function both inside and outside their normal jobs. The primary problem in any consulting arrangement is to ensure there is no conflict of interest for the federal employee. If conflict-of-interest questions can be answered properly, then consulting arrangements can be made that either reimburse the laboratory for work done during normal hours or pay the employee for after-hours consulting.

The Tech affiliates program at NASA's Jet Propulsion Laboratory arranges consulting by its employees to companies that agree to put up a $50,000 retainer before the work begins. Ten thousand dollars covers overhead and at least $65 to $75 per hour is charged for actual technical work. This may be prohibitive for small companies, but more large companies enter consulting agreements every year.

Conflict of Interest Guidelines

When considering a consulting arrangement, federal employees are admonished not to become involved in consulting that would conflict in any way with their laboratory duties or cause them to have an interest in shaping the laboratory's program to favor some particular commercial organization.

Federal employees are also warned against double compensation arrangements. For instance, the employee cannot stall a government contractor's request for information in order to establish a basis for a consulting contract. Other situations in which government employees are not allowed to enter into consulting contracts include:

- No consulting with a firm working under an agency contract in which the employee participates personally and substantially through decision, approval, disapproval, recommendation, the rendering of advice, investigation, or otherwise.

- No consulting with a firm in which the employee will be contributing to that company's efforts on any project sponsored by the agency or being prepared for consideration by the agency.

- No consulting with a firm under consideration for a government contract where the individual can influence the choice of contractor.

- No consulting with a firm that involves methods, techniques, or ideas generated at the agency in the course of normal work until the laboratory's obligation for reporting on that subject has been discharged.

One other principle of consulting not in the written guidelines, but definitely a good business practice, is that the researcher should not be concurrently consulting for a competitor.

THE NUMBER OF FEDERAL RESEARCHERS INVOLVED IN CONSULTING

The federal government's R & D system represents a network of six hundred laboratories and scientific and technical centers that conduct 85 percent of the government's in-house R & D work. These facilities spend more than $20 billion annually and employ more than one hundred thousand scientists and engineers —one-sixth of the nation's total.

For example, in 1987, some 697 Department of Energy (DOE) laboratory employees provided consulting services to industry in fields ranging from advanced materials to application of systems models.

The NREL staff responds to more than three thousand technical inquiries annually. To the degree that resources permit, NREL specialists provide advice, guidance, references, and technical assistance to individuals and organizations. NREL is permitted to accept non-DOE funding for specific research projects and encourages private sector and other agency sponsors when such work relates to, and furthers, the institute's mission.

The U.S. Army's Harry Diamond Laboratories' Technical Volunteer Service (TVS) provides volunteers to assist local government and science and technology educational institutions. This is a public service for which there is no charge. Beyond actual technical problem solving, outreach to the schools motivates students to pursue technical degrees. Many other labs of other agencies encourage similar volunteer efforts.

NASA's Applications Engineering Program

NASA established its Applications Engineering Program (AEP) in 1970 to provide direct NASA assistance and primary funding to promote secondary use of aerospace technology. The AEP involves cooperative efforts to build and test prototype hardware if the industrial partner agrees to provide partial funding and is prepared to commercialize the product. Since its inception, NASA's AEP has initiated more than 150 projects with 75 successful transfers completed.

Unlike a cooperative research and development agreement (CRADA), in which each party must bear its own costs, NASA's AEP provides funds to the commercializing company, although the company must be selected competitively. One way to gain an advantage in the competition might be to first work under a CRADA (in NASA this is called a "Memorandum of Understanding").

Currently, there are at least sixty AEPs at NASA field centers. Over the last ten years, NASA has averaged spending $4 million per year on its AEPs.

One successful AEP was with Bob's Candies of Albany, Georgia. Through a $100,000 shared-cost test project, Bob's Candies is now using heat pipe technology, originally designed to cool our nation's satellites, to control the humidity in warehouses and keep the colorful stripes on candy canes.

Another successful AEP involved the development of the intravascular excimer laser, under exclusive license to Advanced Interventional Systems, Inc., which was funded by grants from NASA and the National Heart, Lung, and Blood Institute.

At the Lewis Research Center, a dual ion beam process for the deposition of diamondlike coatings was developed and transferred to Diamonex, Inc., an Air Products and Chemicals, Inc. subsidiary, as the basis of what should be a tremendous commercial success in the treatment of optical surfaces.

Persons interested in NASA's AEP should first contact the field center Technology Utilization Office, or the Technology Applications Team at Research Triangle Institute (RTI), Research Triangle Park, North Carolina, to define the problem or product development assistance they need.

Small Business Innovation Research (SBIR)

The SBIR program is similar to NASA's AEP, in contrast to CRADAs, in that federal dollars may go to the private sector participant. Data developed under an SBIR can remain proprietary if the government's guidelines are closely followed. An SBIR relationship will not only provide grant money ($50,000 for phase I, up to $500,000 for phase II), but also will create contacts with federal laboratory experts. In addition, the Federal Laboratory Consortium promotes participation by assisting individual laboratories' outreach to small firms and by exhibits and presentations at national SBIR conferences. The Small Business Administration offers the free use of two data bases to help SBIR winners find larger companies or venture capital to exploit their inventions.

Under the SBIR program, U.S. Composites, Inc., of Troy, New York, was awarded a contract by the U.S. Army Materials Technology Laboratory (MTL) to develop a resin applicator ring to continuously impregnate moving fibers with resin in a controlled environment for effective braiding of composites. This project helped eliminate some of private industry's reluctance to invest in an idea from this small entrepreneur by making an effective demonstration of the process. With support provided by MTL, U.S. Composites built a production scale resin applicator ring for composite braiding. Once the ring was successfully tested, U.S. Composites obtained commercialization funding from E. I. DuPont deNemours Company's newly formed composites division as part of a licensing agreement.

Some companies are finding SBIRs are a way to obtain licensing agreements for government technology. For example, Lawrence Livermore National Laboratory is providing temporary loans of unused equipment and technical consultation and advice to a small corporation, which has been awarded a DOE SBIR grant entitled "Advanced Accelerator Development for Industrial Applications." Negotiations are under way to license relevant DOE-patented technology to the firm. The resulting products will be commercially viable accelerators for food treatment and medical equipment sterilization.

Advanced Technology Program (ATP)

The ATP provides technology development grants to U.S. firms or consortia of firms for R & D on precompetitive technologies. Awards to individual firms are limited to $2 million over three years and can be used only for direct R & D costs. Awards to joint ventures can be for up to five years and are limited only by available funds. No direct funding may go to universities or government organizations, although they may participate as members of a joint venture. More than half of the total costs are to be paid by the private sector participants. The ATP will support development of laboratory prototypes and proof of technical feasibility, but not commercial prototypes or proof of commercial feasibility.

In 1991, NIST provided $9 million to eleven projects. The next solicitation is expected to provide $35.9 million.

In Summary

It is now clearly in the interests of the inventive people working as employees or under the sponsorship of the federal government to play active roles in technology transfer. As they become better equipped to fulfill these roles and as enabling changes filter through the government management structure, U. S. industry would be shortsighted if it does not take advantage of the tremendous national resource that the government has funded.

PART V

EPILOGUE

In the fifth year since the passage of the Technology Transfer Act of 1986 it is time to review the promises of the act, some of the apparent shortfalls, and prospects for further progress.

THE PROMISE

The 1986 act did not stand alone. It built on the Stevenson-Wydler Technology Innovation Act of 1980 (P.L. 96-480), which sought to stimulate access to and use of federal technology. The 1980 act had many good intentions but little leverage. Basically, it legitimized technology transfer in federal laboratories where there already was a willingness to undertake such activities.

The disappointing results of the 1980 act led to a considerably stronger 1986 act that was more of a mandate, but it still was much like an authorization act without an appropriation. The 1986 act *did have* the advantage of good timing, enacted when international competitiveness and greater use of technology had become a national issue. The 1986 act promised to strengthen action needed to be globally competitive.

THE SHORTFALL

Five years after the passage of the act, guidance from the agencies still remains incomplete. To some extent this vacuum has been filled by informal guidance at the midmanagement or laboratory level by those individuals who want to see technology transfer succeed. However, it is an indictment of our sluggish bureaucratic system that each department and agency still needs to provide guidance on all points of the legislation.

The 1986 act specifically indicated that position descriptions of scientists and engineers were to be modified to include technology transfer as a legitimate activity in their annual evaluations. Actually, nothing has been done except for some discussions in a few agencies. Implementing regulations or instructions have cascaded down through the bureaucracy. Often they include reference to having technology transfer included in the personnel evaluations of laboratory scientists and engineers. On the other hand, personnel offices have done nothing to change the system.

Another area of slow implementation has been establishing awards at the agency or departmental levels. The drive for recognition has been more evident at the laboratory and bureau level than at the agency level—although there are some notable exceptions.

Sponsors of the 1986 act wanted a more complete reporting mechanism than had been possible under the 1980 act in order to better judge the full array of activities anticipated. Unfortunately, this was translated by the Office of Management and Budget into a rather dry statistical requirement—principally relating to patents awarded, license agreements concluded, and royalty income, rather than information that reflected the extent and nature of technology transfer activities. This has been partially remedied by the National Competitiveness Technology Transfer Act of 1989.

The 1989 amendments made four changes: (1) they brought government-owned, contractor-operated (GOCO) laboratories under the umbrella of the 1986 Technology Transfer Act; (2) they provided protection from disclosure for information developed as part of cooperative research and development agreements (CRADAs) for up to five years—a problem that was causing industry to avoid involvement with CRADAs; (3) they liberalized the funds allotted to technology transfer from R & D programs; and (4) they required an annual report to Congress on both technology transfer activities and plans as part of the annual agency budget submission.

Finally, one must temper criticism about the slow pace of implementation by recognizing that the 1986 act required a nearly complete philosophical reversal of the government R & D community's relationship to industry—from total uninvolvement to full cooperation. This has not only been difficult for agency and laboratory management, but it has also been a challenge for the bench scientists and engineers.

Typically, government scientists and managers have not taken such declarations of a change in climate at face value. They have reservations about whether both agency leadership and Congress are serious about this rather substantial change. Many government researchers have had little or no experience in dealing with their counterparts from industry as colleagues working to meet common technical *and* organizational goals.

The practice of making all unclassified research results freely available to the public cannot be pursued under this changed policy, and it also raises questions of fairness and even potential conflict of interest.

Prospects for the Future

If continuing political support is maintained in Congress, coupled with increased awareness at the political level in the departments and agencies, considerable progress can be anticipated in the future. Successful technology is a long-term investment and requires substantial patience *at all levels of management*.

Officers responsible for technology transfer at various agencies and laboratories identify three remaining key issues. These are: (1) the need to sort out concerns about potential institutional and personal conflicts of interest, (2) the need to remove obstacles to licensing, and (3) the avoidance of burdensome procedures that inhibit the successful negotiation of CRADAs.

Although there is a mounting concern about how best to deal with potential conflict-of-interest situations, there has been little systematic discussion or investigation devoted to this topic. It has been the subject of deliberation by the interagency working group chaired by Department of Commerce officials where agency representatives meet regularly. This area deserves systematic consideration across the government in order to provide suitable guidance to the laboratory scientists and engineers, as well as to avoid undue regulation. As the interaction between the federal research community and private industry increases, the potential for conflict of interest will rise. It will be important for lab directors (as well as for their agency superiors) to navigate a course between overregulation, which would terminate cooperation, and inattention to deficiencies in conduct, which could generate political retribution and the end of cooperative ventures.

Licensing practices probably will be sorted out as the various agencies experiment with procedures to assure timely action, with reasonable approaches to determining value, and with satisfactory means to distribute the proceeds. The one area with which there has been the least experience is that of protecting the value of the license. Virtually all federal patenting accomplished before the 1986 act has been defensive patenting, with little attention given to potential proceeds and the need to protect the *value* of the patent for those to whom it is licensed. It remains to be seen whether the current federal practice of having the Department of Justice defend federal agency interests with respect to patent licensing will be adequate for such protection in the future; or whether it may be possible and desirable to find some other mechanism to enforce this protection adequately.

Discussions with both agency and laboratory technology transfer officials suggest that there remains some confusion about how a CRADA should be put together. In some instances a considerable amount of standard regulation is being borrowed from the procurement process in order to protect the government's interest. This is contrary to the intent of the act which deliberately put the CRADA *outside* the procurement process. Those agencies that have an expedited system for the development and review of CRADAs (such as the National Institutes of Health and the Agricultural Research Service) have not yet experienced serious difficulty in protecting the government's interest. This has been done without attaching a lot of impediments associated with the procurement process. Although model CRADAs have been made available, there probably needs to be greater information exchange from one agency to another—and, especially, to the field—to provide the necessary guidance. Management personnel also need to understand that the procurement process is inappropriate if the CRADA is to meet the requirements of the 1986 act.

In summary, the 1986 act has not proved to be a cure-all for problems of U.S. global competitiveness. It carries some danger in the changed relationship with industry. But it has significantly stimulated the process of opening federal technology to domestic, commercial use. The 1986 act strikes a harmonious chord in promoting cooperation, innovative and entrepreneurial behavior, and reaching out to new opportunities.

APPENDIX I

Technology Transfer Contacts

Listing of Federal Laboratory Technology Transfer Contacts

The following is a list of all U.S. government laboratories and their technology transfer contacts. This list complements the information cited in the main part of this text. For larger laboratories, the contact point listed will be the technology transfer office for the entire laboratory.

Consideration should be exercised in contacting these offices. Although each wants to promote its organization's special technologies and resources to U.S. companies, these offices prefer to be contacted only when a business or researcher has a clear idea of their needs. These offices are NOT information services. Before contacting them, be sure to consult a library or information center to obtain background information on a particular topic.

This list is arranged by agency, then state, then by laboratory name.

Department of Agriculture

Colorado

Denver Wildlife Research Center, APHIS, Federal Center, Bldg. 16, P.O. Box 25266, Denver, CO 80225-0266. Russell Reidinger (303) 236-7820

Department of Agriculture; Agricultural Research Service

Alabama

Animal Parasite Research Unit, P.O. Box 952, Auburn, AL 36830. Phillip H. Klesius (205) 826-4382

National Soil Dynamics Laboratory, P.O. Box 792, Auburn, AL 36831-0792. Jerrel B. Powell (205) 884-4741

Arizona

Arid Watershed Mgt. Research. USDA-ARS, 2000 E. Allen Rd., Tucson, AZ 85719. Frasier Renard (602) 629-6381

Carl Hayden Bee Research Center, 2000 E. Allen Rd., Tucson, AZ 85719. Eric H. Erickson (602) 629-6380

U.S. Water Conservation Laboratory, USDA-ARS-PWA, 4331 E. Broadway Rd., Phoenix, AZ 85040. Herman Bouwer (602) 261-4356

Western Cotton Research Laboratory, USDA-ARS, 4135 E. Broadway, Phoenix, AZ 85040. Thomas J. Henneberry (602) 261-3524

CALIFORNIA

Fruit and Vegetable Chemistry Laboratory, USDA-S & E-ARS, 262 S. Chester Ave., Pasadena, CA 91106. R. M. Horowitz (818) 796-0239

Fruit Breeding and Genetics Research Unit, 2021 South Peach Ave., Fresno, CA 93727. (209) 453-3060

Horticultural Crops Research Laboratory, U.S. Department of Agriculture, ARS, 2021 South Peach Ave., Fresno, CA 93727. Patrick V. Vail (209) 453-3000

Sugarbeet Production Research, U.S. Agricultural Research Sta., 1636 E. Alisal St., Salinas, CA 93915. James E. Duffus (408) 755-2800

U.S. Salinity Laboratory, 4500 Glenwood Dr., Riverside, CA 92501. James D. Rhoades (714) 369-4814

Western Human Nutrition Research Center, P.O. Box 29997, Presidio of San Francisco, San Francisco, CA 94129. James M. Iacono (415) 556-9697

Western Regional Research Center, USDA-ARS-PWA, 800 Buchanan St., Albany, CA 94710. Martin H. Rogoff (415) 559-5600

COLORADO

Crops Research Laboratory, Sugar Beet Production Unit, Colorado State Univ., Ft. Collins, CO 80523. Richard J. Hecker (303) 482-7717

National Seed Storage Laboratory, USDA-ARS, Colorado State Univ., Ft. Collins, CO 80523. Loren Wiesner (303) 484-0402

DELAWARE

Beneficial Insects Research Laboratory, 501 S. Chapel St., Newark, DE 19713. R. J. Dysart (302) 731-7330

FLORIDA

Aquatic Plant Management Laboratory, USDA-ARS-SR-SAA, 3205 College Ave., Ft. Lauderdale, FL 33314. Kerry K. Steward (305) 475-0541

Biological Control of Aquatic Weeds, P.O. Box 1269, Gainesville, FL 32602. Gary R. Buckingham (904) 372-3505 ext. 124

Citrus and Subtropical Products Laboratory, USDA-S & E-ARS-SR, P.O. Box 1909, Winter Haven, FL 33880. Phillip E. Shaw (813) 293-4133

Insect Attractants, Behavior, and Basic Biology Research Laboratory, USDA-ARS, 1700 SW 23rd Dr., P.O. Box 14565, Gainesville, FL 32604. Herbert Oberlander (904) 374-5701

Insects Affecting Man and Animals Research Laboratory, USDA-S & E-ARS-SR, 1600 SW 23rd Dr., P.O. Box 14565, Gainesville, FL 32604. Gary A. Mount (904) 374-5901

Plant Stress and Protection Research Unit, Univ. of Florida, Agronomy Physiology Lab., Bldg. 164, Gainesville, FL 32611. S. H. West (904) 392-1821

Subtropical Horticultural Research Unit, 13601 Old Cutler Rd., Miami, FL 33158. Jennifer L. Sharp (305) 238-9321

U.S. Horticultural Laboratory, USDA-ARS-SR, 2120 Camden Rd., Orlando, FL 32803. Richard T. Mayer (407) 897-7300

GEORGIA

Coastal Plain Experiment Station, USDA-ARS-SR, P.O. Box 748, Tifton, GA 31793. James L. Butler (912) 386-3585

Insect Biology and Population Management Research Laboratory, USDA-ARS, P.O. Box 784, Tifton, GA 31793-0748. Charlie E. Rogers (912) 387-2320

National Peanut Research Laboratory, 1011 Forrester Dr., SE, Dawson, GA 31742. Richard J. Cole (912) 995-4441

Nematodes, Weeds and Crops Research, Georgia Coastal Plain Exp. Sta., Tifton, GA 31793. Alva W. Johnson (912) 386-3372

Richard B. Russell Agriculture Research Center, USDA-ARS-SAA, P.O. Box 5677, Athens, GA 30613. David Zimmer (404) 546-3541

Southeast Poultry Research Laboratory, USDA-ARS-SAA, 934 College Station Rd., P.O. Box 5657, Athens, GA 30605. Charles W. Beard (404) 546-3434

Southeast Watershed Research Laboratory, P.O. Box 946, Tifton, GA 31793. Ralph Leonard (912) 386-3462

Southeastern Fruit and Tree Nut Laboratory, USDA-ARS-SAA, P.O. Box 87, Byron, GA 31008. James W. Snow (912) 956-5656

Southern Piedmont Conservation Laboratory, USDA-S & E-ARS-SR, Highway 53, P.O. Box 555, Watkinsville, GA 30677. Maurice H. Frere (404) 769-5631

Stored-Product Insects Research and Development Laboratory, USDA-ARS-SAA, P.O. Box 22909, Savannah, GA 31403. Robert Davis (912) 233-7981

HAWAII

Tropical Fruit and Vegetable Research Laboratory, USDA-ARS-PWA, 2727 Woodlawn Dr., P.O. Box 2280, Honolulu, HI 96804. J. E. Gilmore (808) 988-2158

IDAHO

Soil and Water Management Research Unit, 3793 North 3600 East, Kimberly, ID 83341. David L. Carter (208) 423-5582

ILLINOIS

Northern Regional Research Center, USDA-S & E-ARS-MWA, 1815 N. University St., Peoria, IL 61604. L. H. Princen (309) 685-4011

INDIANA

National Soil Erosion Laboratory, USDA-S & E-ARS, Purdue Univ., Soil Bldg., West Lafayette, IN 47907. John M. Laflen (317) 494-8673

IOWA

Corn Insects Research Unit, P.O. Box 45B, R. Route 3, Ankeny, IA 50021. Wilbur D. Guthrie (515) 964-6664

National Animal Disease Center, USDA-ARS-CPA, P.O. Box 70, Ames, IA 50010. Harley W. Moon (515) 239-8201

KANSAS

U.S. Grain Marketing Research Laboratory, USDA-ARS-NPA, 1515 College Ave., Manhattan, KS 66502. R. L. Dunkle (913) 776-2701

KENTUCKY

Tobacco and Forage Research, Rm. 107-A, Animal Pathology Bldg., USDA-ARS, Lexington, KY 40546-0076. Steven J. Crafts-Brandner (606) 257-4770

LOUISIANA

Honey Bee Breeding, Genetics, and Physiology Research Laboratory, 1157 Ben Hur Rd., Baton Rouge, LA 70820. Thomas E. Rinderer (504) 766-6064

Southern Regional Research Center, USDA-S & E-ARS-SRRC, P.O. Box 19687, 1100 Robert E. Lee Blvd., New Orleans, LA 70179. John A. Barkate (504) 286-4212

U.S. Sugarcane Field Laboratory, USDA-ARS-Mid South Area, P.O. Box 470, Houma, LA 70361. Rex W. Millhollon (504) 872-5042

MAINE

New England Plant, Soil and Water Laboratory, Univ. of Maine, Orono, ME 04469. William Clapham (207) 581-3266

MARYLAND

Beltsville Agricultural Research Center, Technology Transfer Coordinator, Bldg. 005, Rm. 404, Beltsville, MD 20705. James Hall (301) 344-4045

Beltsville Human Nutrition Research Center, USDA-ARS-NER, Rm. 223, Bldg. 308, BARC-E, Beltsville, MD 20705. Helene N. Guttman (301) 344-1627

Foreign Disease-Weed Science Resea. Agricultural Research Ser., U.S. Department of Agriculture, Ft. Detrick, Bldg. 1301, Fredrick, MD 21701. William M. Dowler (301) 663-7344

MASSACHUSETTS

USDA Human Nutrition Research Center on Aging at Tufts Univ., Tufts Univ., 711 Washington St., Boston, MA 02111. Ronald L. Prior (617) 556-3310

MICHIGAN

Regional Poultry Research Laboratory, USDA-ARS-MWA, 3606 East Mt. Hope Rd., East Lansing, MI 48823. Richard L. Witter (517) 337-6828

MINNESOTA

Cereal Rust Laboratory, Univ. of Minnesota, 1551 Lindig Ave., St. Paul, MN 55108. K. J. Leonard (612) 376-4647

North Central Soil Conservation Research Laboratory, USDA-ARS, North Iowa Ave., Morris, MN 56267. Charles A. Onstad (612) 589-3411

Plant Science Research Unit, Univ. of Minnesota, 411 Borlaug Hall, 1991 Upper Buford Cir., St. Paul, MN 55108. Gordon C. Marten (612) 373-7281

Red River Valley Potato Research Laboratory, 311 Fifth Ave., NE, P.O. Box 113, East Grand Forks, MN 56721. Paul H. Orr (218) 773-2473

MISSISSIPPI

Boll Weevil Research Unit, USDA-ARS, P.O. Box 5367, MS State, MS 39762. James Smith (601) 323-2230

Cotton Physiology and Genetics Research Laboratory, Delta States Research Center, P.O. Box 345, Stoneville, MS 38776. William R. Meredith (601) 686-2311 ext. 241

Crop Science Research Laboratory, USDA-ARS-SR, P.O. Box 5367, Mississippi State, MS 39762. Johnie N. Jenkins (601) 323-2230

Field Crops Mechanization Research Unit, P.O. Box 36, Stoneville, MS 38776. J. Ray Williford (601) 686-2311 ext. 352

National Monitoring and Residue Analysis Lab., 3505 25th Ave., P.O. Box 3209, Gulfport, MS 39505-3209. Joseph H. Ford (601) 863-8124

National Sedimentation Laboratory, P.O. Box 1157, Airport Rd., Oxford, MS 38655. C. K. Mutchler (601) 232-2900

South Central Poultry Research Laboratory, P.O. Box 5367, Mississippi State, MS 39762. James W. Deaton (601) 323-2230

Southern Field Crop Insect Management Laboratory, Delta States Research Center, P.O. Box 346, Stoneville, MS 38776. D. D. Hardee (601) 686-2311 ext. 231

Southern Weed Science Laboratory, USDA-ARS, P.O. Box 350, Stoneville, MS 38776. Stephen O. Duke (601) 686-2311

Soybean Production Research Unit, P.O. Box 196, Stoneville, MS 38776. Thomas C. Kilen (601) 686-9311 ext. 232

U.S. Cotton Ginning Laboratory, P.O. Box 256, Stoneville, MS 38776. W. Stanley Anthony (601) 686-2385

MISSOURI

Biological Control of Insects Research, USDA-S & E-ARS-NCR, P.O. Box 7629, Research Park, Columbia, MO 65205. Carlo M. Ignoffo (314) 875-5361

Cropping Systems and Water Quality Research Unit, 207 Business Loop 70 East, Columbia, MO 65211. Maurice R. Gebhardt (314) 882-1114

NEBRASKA

Livestock Insects Research Laboratory, Univ. of Nebraska, E. Campus, 305A Plant Industry Bldg., Lincoln, NE 68583. G. D. Thomas (402) 471-5261

Roman L. Hruska U.S. Meat Animal Research Center, USDA-ARS-CPA, State Spur 18D, P.O. Box 166, Clay Center, NE 68933. D. B. Laster (402) 762-4109

NEW MEXICO

Southwestern Cotton Ginning Research Laboratory, P.O. Box 578, 300 E. College Dr., Mesilla Park, NM 88047. Sidney E. Hughs (505) 526-6381

NEW YORK

Germplasm Resources/USDA-ARS, New York St. Agr. Exp. Sta., Geneva, NY 14456-0462. Stephen Kresovich (315) 787-2244

Plum Island Animal Disease Center, USDA-ARS, P.O. Box 848, Greenport, NY 11944. Roger G. Breeze (516) 323-2500

U.S. Plant, Soil and Nutrition Laboratory, Cornell Univ., Tower Rd., Ithaca, NY 14853. D. R. Van Campen (607) 256-5480

NORTH CAROLINA

Crops Research Laboratory, USDA-ARS-SAA, P.O. Box 1555, Oxford, NC 27565-1555. Harvey W. Spurr, Jr. (919) 693-5151

Plant Physiology and Photosynthesis Research Unit, North Carolina State Univ., P.O. Box 7620, Raleigh, NC 27695. Donald E. Moreland (919) 737-2661

Soybean and Nitrogen Fixation Research Unit, North Carolina State Univ., P.O. Box 7620, 4114 Williams Hall, Raleigh, NC 27695-7620. Richard F. Wilson (919) 737-3171

NORTH DAKOTA

Human Nutrition Research Center, 2420 2nd Ave., N, P.O. Box 7166, Univ. Station, Grand Forks, ND 58202. Forrest H. Nielson (701) 795-8353

Northern Great Plains Research Center, USDA-S & E-ARS-NPA, P.O. Box 459, Mandan, ND 58554. Albert B. Frank (701) 663-6445

Red River Valley Agricultural Research Center, USDA-ARS, P.O. Box 5677, University Station, Fargo, ND 58105. Don C. Zimmerman (701) 239-1370

OHIO

North Appalachian Experimental Watershed, Research Leader, P.O. Box 478, Coshocton, OH 43812. L. B. Owens (614) 545-6349

Ohio Agricultural Research and Development Center, 1680 Madison Ave., Wooster, OH 44691-4096. Robert Furbee (216) 263-3777

U.S. National Arboretum, 359 Main Rd., Delaware, OH 43015. Lawrence Schreiber (614) 363-1129

OKLAHOMA

Forage and Livestock Research Labor, USDA-ARS, P.O. Box 1199, El Reno, OK 73036. William A. Phillips (405) 262-5291

Plant Science and Water Conservation, Plant Science Research Lab., 1301 N. Western, Stillwater, OK 74075. Robert L. Burton (405) 624-4126

Water Quality and Watershed Research Laboratory, P.O. Box 1430, Durant, OK 74702. S. J. Smith (405) 924-5066

OREGON

Horticultural Crops Research Laboratory, 3420 Northwest Orchard Ave., Corvallis, OR 97330. Robert G. Linderman (503) 757-4544

National Clonal Germplasm Repository, 33447 Peoria Rd., Corvallis, OR 97333. Kim Hummer (503) 757-4448

Pennsylvania

Eastern Regional Research Center, USDA-S & E-ARS-NAA, 600 E. Mermaid La., Philadelphia, PA 19118. John P. Cherry (215) 233-6595

Northeast Watershed Research Laboratory, 111 Research Bldg. A, University Park, PA 16802. Harry B. Pionke (814) 865-2048

U.S. Regional Pasture Research Laboratory, Curtin Rd., University Park, PA 16802. R. R. Hill (814) 863-0939

Puerto Rico

Tropical Agriculture Research Station, Tropical Crops and Germplasm Research Unit, P.O. Box 70, Mayaguez, PR 00709. Antonio Sotomayor-Rios (809) 831-3435

South Carolina

Cotton Production Research Unit, USDA-ARS, P.O. Box 2131, Florence, SC 29503. P. G. Hunt (803) 669-5203

Cotton Quality Research Station, USDA-ARS, P.O. Box 792, Clemson, SC 29631. C. Kenneth Bragg (803) 656-2488

Soil and Water Conservation Research Unit, Darlington Hwy., P.O. Box 3039, Florence, SC 29502. Patrick G. Hunt (803) 669-5203

U.S. Vegetable Laboratory, USDA-ARS-SAA, 2875 Savannah Hwy., Charleston, SC 29414. George Fassuliotis (803) 556-0840

South Dakota

Northern Grain Insects Research Laboratory, USDA-S & E-ARS-NCR, R.R. #3, Brookings, SD 57006. Gerald R. Sutter (605) 693-5201

Texas

Children's Nutrition Research Center, Department of Agriculture, Baylor College of Medicine, 1709 Dryden, Suite 601, Houston, TX 77030. Buford L. Nichols (713) 799-6006

Conservation and Production Research Laboratory, USDA-ARS, P.O. Drawer Box 10, Bushland, TX 79012. B. A. Stewart (806) 378-5724

Cropping Systems Research Laboratory, Route 3, Box 215, Lubbock, TX 79401. Jerry E. Quisenberry (806) 746-5353

Grassland, Soil and Water Research Laboratory, USDA-ARS, 808 East Blackland Rd., Temple, TX 76502. Clarence W. Richardson (817) 770-6500

Knipling-Bushland U.S. Livestock Insects Laboratory, P.O. Box 232, Kerrville, TX 78029-0232. S. E. Kunz (512) 257-3566

Rice Research Unit, Route 7, Box 999, Beaumont, TX 77706. Charles N. Bollich (409) 752-5221

Subtropical Agricultural Research Laboratory, 2301 S. International Blvd., Weslaco, TX 78596. Edgar G. King (512) 565-2606

Veterinary Toxicology and Entomology Laboratory, Route 5, P.O. Box 810, College Station, TX 77840. G. Wayne Ivie (409) 260-9372

Utah

Poisonous Plant Research Laboratory, Agricultural Research Service, 1150 E. 1400 North, Logan UT 84321. Lynn F. James (801) 752-2941

Washington

Fruit and Vegetable Insect Research Unit, 3706 W. Nob Hill Blvd., Yakima, WA 98902. J. L. Krysan (509) 575-5900

Fruit Research Laboratory, 1104 N. Western Ave., Wenatchee, WA 98801. Max W. Williams (509) 664-2280

Yakima Agricultural Research Lab., USDA-ARS, Pacific West Area, 3706 W. Nob Hill Blvd., Yakima, WA 98902. J. L. Krysan (509) 575-5877

WEST VIRGINIA

Appalachian Soil and Water Conservation Research Laboratory, P.O. Box 867, Beckley, WV 25801. (304) 252-6426

Appalachian Fruit Research Station, USDA-ARS-NER, Route 2, Box 45, Kearneysville, WV 25430. Bill A. Butt (304) 725-3451

WISCONSIN

Cereal Crops Research Unit, ARS, 501 N. Walnut St., Madison, WI 53705. David M. Peterson (608) 262-4482

U.S. Dairy Forage Research Center, 1925 Linden Dr. W., Univ. of Wisconsin, Madison, WI 53706. Larry D. Satter (608) 263-2030

WYOMING

Athropod-Borne Animal Diseases Research Laboratory, USDA-S & E-ARS, P.O. Box 3965, University Station, Laramie, WY 82071. Thomas E. Walton (307) 721-0304

DEPARTMENT OF AGRICULTURE; FOREST SERVICE

CALIFORNIA

Pacific Southwest Forest and Range Station, P.O. Box 245, Berkeley, CA 94701. Richard Hubberd (415) 486-3286

COLORADO

Rocky Mountain Forest and Range Experiment Station, 240 W. Prospect Rd., Ft. Collins, CO 80526. Robert Hamre (303) 498-1282

LOUISIANA

Southern Forest Experimental Station, Rm. T-10210, 701 Loyola Ave., New Orleans, LA 70113. James H. Perdue (504) 589-6712

MINNESOTA

North Central Forest Experimental Station, 1992 Folwell Ave., St. Paul, MN 55108. Nancy R. Walters (612) 642-5252

NORTH CAROLINA

Southeastern Forest Experimental Station, Forest Service, USDA, P.O. Box 2680, Asheville, NC 28802. Gordon Lewis (704) 257-4304

OREGON

Pacific Northwest Research Station, P.O. Box 3890, Portland, OR 97208. Cynthia Miner (503) 326-7135

PENNSYLVANIA

Northeastern Forest Experimental Station, 370 Reed Rd., Broomall, PA 19008. Robert Lewis (215) 690-3048

UTAH

Intermountain Research Station, P & A AD, Deputy Director, 324 25th St., Ogden, UT 84401. Carter B. Gibbs (801) 625-5412

WISCONSIN

Forest Products Laboratory, One Gifford Pinchot Dr., Madison, WI 53705-2398. Rodney G. Larson (608) 231-9200

DEPARTMENT OF COMMERCE

MARYLAND

National Institute of Standards and Technology, Office for Technology Utilization, Bldg. 101, Rm. 537, Gaithersburg, MD 20899. David Edgerly (301) 975-3087

DEPARTMENT OF COMMERCE;
NATIONAL OCEANIC
AND ATMOSPHERIC
ADMINISTRATION (NOAA)

ALASKA

Auke Bay Laboratory, National Marine Fisheries Service, P.O. Box 210155, Auke Bay, AK 99821. George R. Snyder (907) 789-7231

CALIFORNIA

Pacific Environmental Group, National Marine Fisheries Service, P.O. Box 831, Monterey, CA 93942. Andrew Bakun (408) 646-3311

Southwest Fisheries Center and La Jolla Fisheries Laboratory, National Marine Fisheries Service, Box 271, La Jolla, CA 92038. Izadore Barrett (619) 453-2820

Tiburon Laboratory, National Marine Fisheries Service, 3150 Paradise Dr., Tiburon, CA 94920. Norman J. Abramson (415) 435-3149

COLORADO

Aeronomy Laboratory, Environmental Research Laboratories, 325 Broadway, Boulder, CO 80303. Daniel L. Albritten (303) 497-3218

Environmental Research Laboratories, Office of Oceanic & Atmospheric Research, R/E5X2, 325 Broadway, Boulder, CO 80303. B. L. Trotter (303) 497-6914

Environmental Sciences Group, Environmental Research Labs., RE2, 3100 Marine St., Boulder, CO 80303. William Hooke (303) 497-6378

National Geophysical Data Center, NOAA/NESDIS, 325 Broadway, Boulder, CO 80303. Michael Chinnery (303) 497-6215

Space Environment Laboratory, Environmental Research Laboratories, 325 Broadway, Boulder, CO 80303. Ernest Hildner (303) 497-3311

Wave Propagation Laboratory, Environmental Research Laboratories, R/E/WP 325 Broadway, Boulder, CO 80303. Steven Clifford (303) 497-6261

CONNECTICUT

Milford Laboratory, National Marine Fisheries Service, Milford, CT 06460. Anthony Calabrese (203) 878-2459

DISTRICT OF COLUMBIA

National Meteorological Center, National Weather Service, Development Division, Rm. 204, WWB, Washington, DC 20233. John A. Brown (301) 763-8005

National Meteorological Center, National Weather Service, Climate Analysis Center, Rm. 606 WWB, Washington, DC 20233. David R. Rodenhuis (301) 763-8167

National Systematics Laboratory, National Marine Fisheries Service, Smithsonian Institution, Washington, DC 20560. Bruce B. Collette (202) 357-2550

Satellite Applications Laboratory, NESDIS, Rm. 601, WWB, Washington, DC 20233. Donald B. Miller (301) 763-8282

Satellite Research Laboratory, NESDIS, WWB/Rm. 712, Washington, DC 20233. George Ohring (301) 763-8078

FLORIDA

Atlantic Oceanographic and Meteorological Laboratories, Environmental Research Laboratories, 4301 Rickenbacker Causeway, Miami, FL 33149. Hugo F. Bezdek (305) 361-4300

Panama City Laboratory, National Marine Fisheries Service, Panama City, FL 32407-7499. Eugene L. Nakamura (904) 234-6541

Southeast Fisheries Center and Miami Fisheries Laboratory, National Marine Fisheries Service, 75 Virginia Beach Dr., Miami, FL 33149. Brad Brown (305) 361-4286

Hawaii

Honolulu Laboratory, National Marine Fisheries Service, P.O. Box 3830, Honolulu, HI 96812. Richard Shomura (808) 943-1221

Maryland

Air Resources Laboratory, Environmental Research Labs, Rm. 927, 8060 13th St., Silver Spring, MD 20910. Lester Machta (301) 427-7684

Hydrologic Research Laboratory, National Weather Service, Rm. 530, 8060 13th St. Silver Spring, MD 20910. Danny L. Freed (301) 427-7619

Integrated Systems Laboratory, National Weather Service, Rm. 201, 8060 13th St., Silver Spring, MD 20910. Donald T. Acheson (301) 427-7809

NOAA Office of Climatic and Atmospheric Research, Office of Oceanic & Atmospheric Research, WSC5, Rm. 825, Rockville, MD 20852. Michael Hall (301) 443-8415

Oxford Laboratory, National Marine Fisheries Service, Oxford, MD 21654. Aaron Rosenfield (301) 226-5193

Techniques Development Laboratory, National Weather Service, Rm. 825, Gramax Bldg. 8060 13th St., Silver Spring, MD 20910. Harry R. Glahn (301) 427-7768

Massachusetts

Gloucester Laboratory, National Marine Fisheries Service, Emerson Ave., Gloucester, MA 01930. Robert J. Learson (617) 281-3600 ext. 237

Northeast Fisheries Center and Woods Hole Fisheries Laboratory, National Marine Fisheries Service, Woods Hole, MA 02543. Allen Peterson (617) 548-5123

Michigan

Great Lakes Environmental Research Laboratory, Environmental Research Labs., 2205 Commonwealth Blvd., Ann Arbor, MI 48105-1593. Alfred M. Beeton (313) 668-2244

Mississippi

Mississippi Laboratory, National Marine Fisheries Service, 3209 Frederick St., P.O. Drawer 1207, Pascagoula, MS 39567. Andrew J. Kemmerer (601) 688-3650

National Data Bouy Center, National Weather Service, NSTL Station, MS 39529. Jerry McCall (601) 688-2822

New Jersey

Geophysical Fluid Dynamics Laboratory, Environmental Research Laboratories, P.O. Box 308, Princeton, NJ 08540. Jerry D. Mahlman (609) 452-6502

Sandy Hook Laboratory, National Marine Fisheries Service, Highlands, NJ 07732. Ann Studholme (201) 872-0200

North Carolina

Beaufort Laboratory, National Marine Fisheries Service, Beaufort, NC 28516-9722. Ford Cross (919) 728-4595

National Climatic Data Center, NESDIS, Federal Bldg., Asheville, NC 28801. Kenneth Hadeen (704) 259-0682

Oklahoma

National Severe Storms Laboratory, Environmental Research Laboratories, 1313 Halley Circle, Norman, OK 73069. Robert A. Maddox (405) 366-0429

Rhode Island

Atlantic Environmental Group, National Marine Fisheries Service, R.R. 7A, Box 522A, Narragansett, RI 02882. Merton C. Ingham (401) 789-9326

Narragansett Laboratory, National Marine Fisheries Service, South Ferry Rd., Narragansett, RI 02882. Kenneth Sherman (401) 789-9326

South Carolina

Charleston Laboratory, National Marine Fisheries Service, P.O. Box 12607, 217 Ft. Johnson Rd., James Island, Charleston, SC 29412. Robert Kifer (803) 762-1200

Texas

Galveston Laboratory, National Marine Fisheries Service, 4700 Avenue U, Galveston, TX 77550. Edward F. Klima (409) 766-3500

Virginia

Equipment Test and Evaluation Branch, National Weather Service, R.D. 1, Box 105, Sterling, VA 22170. Robert Strickler (703) 471-5302

Washington

Northwest and Alaska Fisheries Center and Associated Laboratories, National Marine Fisheries Service, 7600 Sand Point Way, NE, Seattle, WA 98112-0070. George Tananaka (206) 526-4760

Pacific Marine Environmental Laboratory, Environmental Research Labs., 7600 Sand Point Way NE, Bin C15700, Seattle, WA 98115. Eddie L. Bernard (206) 526-6239

Department of Commerce; National Telecommunications and Information Administration

Colorado

Forecast Systems Laboratory, 3100 Marine St., Boulder, CO 80303. A. E. MacDonald (303) 497-6378

Institute for Telecommunication Sciences, 325 Broadway, Boulder, CO 80303-3328. Val O'Day (303) 497-3484

Department of Defense

Maryland

Defense Nuclear Agency, National Naval Medical Center, Bethesda, MD 20814-5145. Director (202) 295-1210

Department of Defense; Air Force

California

Air Force Astronautics Laboratory, AFAL/TSTR, Edwards AFB, CA 93523-5000. Chris Degnan (805) 275-5014

Air Force Flight Test Center, Edwards AFB, CA 93523. William T. Twinting (805) 277-2410

Air Force Rocket Propulsion Laboratory, STINFO Office, AFRPL/TSPR, Edwards AFB, CA 93523. Rosemary Degnan (805) 277-5677

Colorado

Frank J. Seiler Research Laboratory, Chief Scientist Office, FJSRL,NA, USAF Academy, CO 80840. Ken Siegenthaler (303) 472-3120

FLORIDA

Air Force Armament Laboratory, AFATL/DOIR, Eglin AFB, FL 32542-5434. James E. Krug (904) 882-4013

Air Force Engineering and Services Center, AFESC/RD, Tyndall AFB, FL 32403-6001. Bob Van Orman (904) 283-6494

MASSACHUSETTS

Air Force Geophysics Laboratory, AFGL/XO, Hanscom AFB, MA 01731-5000. Rene Cormier (617) 861-3606

NEW MEXICO

Air Force Weapons Laboratory, AFWL/CA, Kirtland AFB, NM 87117-6008. Patrick Rodriquez (505) 844-9856

NEW YORK

Rome Air Development Center, Code RADC-XP, Griffiss AFB, NY 13441-5700. Billy G. Oaks (315) 330-3705

OHIO

Air Force Wright Research and Development Center, ORTA, WRDC/XO, Rm. 219, Wright-Patterson AFB, OH 45433-6523. Cindy Ingalls (513) 255-2788

Harry G. Armstrong Aerospace Medical Research Laboratory, AAMRL/TID, Wright-Patterson AFB, OH 45433. Patricia Lewandowski (513) 255-2423

TENNESSEE

Air Force Arnold Engineering Development Center, Arnold Air Force Station, TN 37389-5000. Dale F. Vosika (615) 454-7621

TEXAS

Air Force Human Resources Laboratory, AFHRL/PRT, IR & D, Brooks AFB, TX 78235-5601. Douglas Blair (512) 536-3426

Air Force School of Aerospace Medicine, USAFSAM/TSZ, Brooks AFB, TX 78235. Patricia Wilson (512) 536-3836

Human Systems Divisions, Technical Plans & Analysis Officer, DN/CC-TT, Brooks AFB, TX 78235-5000. Mark Loper (512) 536-3405

DEPARTMENT OF DEFENSE; ARMY

ALABAMA

Army Aeromedical Research Laboratory, ATTN: SGRD-UAX-SI, P.O. Box 577, Ft. Rucker, AL 36362-5292. Diana Hemphill (205) 255-6907

Army Missile Command, ATTN: AMSMI-RD-TI, Redstone Arsenal, AL 35898-5243. Steven Smith (205) 876-5449

CALIFORNIA

Army Aviation Engineering Flight Activity, SAVTE-P, Edwards AFB, CA 93523-5000. John T. Blaha (805) 277-4643

Letterman Army Institute of Research, SGRD-ULZ-IR, Presidio of San Francisco, CA 94129-6800. Jack Keller (415) 561-2641

DISTRICT OF COLUMBIA

Army Institute of Dental Research, Walter Reed Army Medical Center, ATTN: SGRD-UDR, Bldg. 40, Washington, DC 20307-5300. Gino C. Battistone (202) 576-3254

Army Walter Reed Institute of Research, Attn: SGRD-UWZ-I, Washington, DC 20307-5100. B. Nolan Dale (202) 576-3814

ILLINOIS

Army Construction Engineering Research Laboratory, ATTN: CERER-TAO, P.O.

Box 4005, Champaign, IL 61824-4005. Rob Gorham (217) 373-6789

MARYLAND

Army Ballistic Research Laboratory, ATTN: SLCBR-D, Bldg. 328, Aberdeen Proving Ground, MD 21005-5066. Richard Dimmick (301) 278-6955

Army Biomedical Research and Development Laboratory, ATTN: SGRD-UBZ-C, Bldg. 568, Ft. Detrick, Frederick, MD 21701-5010. Lee Merrell (301) 663-2024

Army Chemical Research, Development, and Engineering Center, ATTN: SMCCR-OPP, Aberdeen Proving Ground, MD 21010-5423. Susan Luckan (301) 671-2031

Army Combat Systems Test Activity, ATTN: STECS, Bldg. 400, Aberdeen Proving Ground, MD 21005-5059. Palmer Paules (301) 274-4102

Army Human Engineering Laboratory, ATTN: SLCHE-SS-IR, Aberdeen Proving Ground, MD 21005-5001. Dean Westerman (301) 278-5817

Army Medical Research Institute of Chemical Defense, ATTN: SGRD-UV-R, Aberdeen Proving Ground, MD 21010-5425. Lloyd Roberts (301) 671-2363

Army Medical Research Institute of Infectious Diseases, Deputy for Science, ATTN: SGRD-UIZ-D, Bldg. 1425, Ft. Detrick, Frederick, MD 21701-5011. Michael A. Chirigos (301) 663-2227

Harry Diamond Laboratories, ATTN: SLCHD-PO-P, 2800 Powder Mill Rd., Adelphi, MD 20783-1197. Mary S. Binseel (301) 394-2952

MASSACHUSETTS

Army Materials Technology Laboratory, ATTN: SLCMT-DA, Watertown, MA 02172-0001. George R. Thomas (617) 923-5527

Army Natick RD & E Center, ATTN: STRNC-EMP, Natick, MA 01760-5014. Robert Rosenkrans (508) 651-5296

Army Research Institute of Environmental Medicine, Bldg. 42, ATTN: SGRD-UE-RSI, Natick, MA 01760-5007. Carol Joriman (617) 651-4891

MICHIGAN

Army Tank-Automotive Center, ATTN: AMSTA-CK, Warren, MI 48397-5000. Robert Hostetler (313) 574-5270

MISSISSIPPI

Army Engineer Waterways Experiment Station, ATTN: CEWES-FV, P.O. Box 631, Vicksburg, MS 39181-0631. Philip Stewart (601) 634-4113

MISSOURI

Army Aviation Systems Command, ATTN: AMSAV-NR, 4300 Goodfellow Blvd., St. Louis, MO 63120-1798. Roy J. Warhover (314) 263-1082

NEW HAMPSHIRE

Army Cold Regions Research & Engineering Laboratory, ATTN: CRREL-CS, 72 Lyme Rd., Hanover, NH 03755-1920. Nancy Liston (603) 646-4221

NEW JERSEY

Army Armament Research, Development and Engineering Center, ATTN: SMCAR-AST, Bldg. 1, Picatinny Arsenal, NJ 07806-5000. Robert Zanowicz (201) 724-7954

Army Communications and Electronics Command, ATTN: AMSEL-RD-TPPO-L, Ft. Monmouth, NJ 07703-5205. Albert J. Feddeler (201) 544-2239

Army Electronics Technology and Devices Laboratory, SLCET-DT, Ft. Monmouth, NJ 07703-5000. Richard Stern (201) 544-4666

New Mexico

Army Atmospheric Sciences Laboratory, LABCOM, ATTN: DELAS-D, White Sands, NM 88002. Carl Wright (505) 678-3504

Army Vulnerability Assessment Laboratory, ATTN: SLCVA-DPC, White Sands Missile Range, NM 88002-5513. Tom Reader (505) 678-2650

North Carolina

Army Research Office, ATTN: SLCRO-TS, P.O. Box 12211, Research Triangle Park, NC 27709-2211. David W. Seitz (919) 549-0641

Texas

Army Institute of Surgical Research, ATTN: SGRD-USA, Ft. Sam Houston, San Antonio, TX 78234-6200. David Howard (512) 221-2340

Utah

Army Dugway Proving Grounds, STEDP-SD, Dugway, UT 84022. Lothar L. Salomon (801) 522-3314

Virginia

Army Belvoir R & D Center, ATTN: STRBE-IL, Ft. Belvoir, VA 22060-5606. Connie Harrisson (703) 664-1068

Army Center for Signals Warfare Center, Communications-Electronics Command, Vint Hill Farms Station, Warrenton, VA 22186-5100. G. William Mitchell (703) 347-6464

Army Engineer Topographic Laboratories, ATTN: CEETL-RT Ft. Belvoir, VA 22060-5546. George N. Simcox (703) 355-2629

Army Night Vision and Electro-Optics Laboratory, ATTN: AMSEL-RD-NV-D, Public Affairs Office, Ft. Belvoir, VA 22060-5677. Clarence Johnson (703) 664-5308

Army Research Institute for Behavioral and Social Sciences, ATTN: PERI-POT, 5001 Eisenhower Ave., Alexandria, VA 22333-5600. Jerry Bialecki (202) 274-8653

Aviation Applied Technology Directorate, USAARTA (AVSCOM), Ft. Eustis, VA 23604-5577. John L. Shipley (804) 878-2000

Department of Defense; Navy

California

Laboratory of Biomedical and Environmental Science, UCLA, 900 Veteran Ave., Los Angeles, CA 90024-1786. William J. Moffitt (213) 825-9431

Naval Biosciences Laboratory, Naval Supply Center, Code 82, Oakland, CA 95624. Richard C. Hedstrom (415) 466-5955

Naval Civil Engineering Laboratory, LO3C, Port Hueneme, CA 93043. Jerry Dummer (805) 982-4070

Naval Health Research Center, Technology Transfer Office, P.O. Box 85122, San Diego, CA 92138-9174. (619) 225-7396

Naval Ocean Systems Center, Code 0141, San Diego, CA 92152-5000. Richard November (619) 553-2103

Naval Weapons Center, Industrial and Governmental Liaison Office, Code 374, China Lake, CA 93555-6001. George F. Linsteadt (619) 939-1074

Navy Personnel R & D Center, Technical Information Office, Code 232, San Diego, CA 92152-6800. Robert Turney (619) 553-9308

Pacific Missile Test Center, Code 1032, Technology Transfer Officer, Point Mugu, CA 93042. Dan Kimsey (805) 989-7124

Connecticut

Naval Submarine Medical Research Laboratory, Box 900, Naval Submarine

Base, Groton, CT 06349-5900.
S. M. Luria (203) 449-3398

Naval Underwater Systems Center, Office of Research and Technology Application, Code 105, Bldg. 80, New London, CT 06320. Margaret M. McNamara (203) 440-4116

DISTRICT OF COLUMBIA

Naval Research Laboratory, Code 1003.2, 4555 Overlook Ave., Washington, DC 20375-5000. George Abraham (202) 767-3521

FLORIDA

Naval Aerospace Medical Research Laboratory, Code 00B2, Naval Air Station, Pensacola, FL 32508-5700. Kathleen Mayer (904) 452-3286

Naval Coastal Systems Center, Code 710T, Panama City, FL 32407-5000. J. D. Wright (904) 235-5275

Naval Training Systems Center, 12350 Research Pkwy., Orlando, FL 32826-3224. H. C. Okraski (407) 380-8135

ILLINOIS

Naval Dental Research Institute, Domestic Technology Transfer, Naval Training Center, Bldg. 1-H, Great Lakes, IL 60088-5259. (312) 688-6782

INDIANA

Naval Avionics Center, Office of Res. and Tech. Appl., Code 802, 6000 E. 21st St., Indianapolis, IN 46219-2189. Larry Halbig (317) 353-7075

Naval Weapons Support Center, Code DPM2, Crane, IN 47522-5000. Bill Chancellor (812) 854-1162

MARYLAND

David Taylor Research Center, Code 0117, Bethesda, MD 20084-5000. Basil Nakonechny (202) 227-1037

Naval Air Test Center, Patuxent River, MD 20870-5304. Dick Gallant (301) 862-7578

Naval Explosive Ordnance Disposal Technology Center, Indian Head, MD 20640-5070. Bert Stevenson (301) 743-4430

Naval Medical Research Institute, ATTN: Technology Transfer Office, Bethesda, MD 20014. L. Kiesow (202) 295-1310

Naval Surface Weapons Center, White Oak, Code D211, Silver Spring, MD 20903-5000. Ramsey D. Johnson (301) 394-1505

MASSACHUSETTS

Naval Blood Research Laboratory, Boston Univ. School of Medicine, 615 Albany Ave., Boston, MA 02118. (617) 638-4950

Navy Clothing and Textile Research Facility, 21 Strathmore Rd., Natick, MA 01760-2490. John Mylotte (508) 651-4680

MISSISSIPPI

Naval Oceanographic and Atmospheric Research Laboratory (NOARL), Code 115T, NSTL Station, Bay St. Louis, MS 39529. George E. Stanford (601) 688-4790

Naval Oceanographic Office, Code PF, Stennis Space Center, Bay St. Louis, MS 39522-5001. James A. Stone (601) 688-5928

NEW JERSEY

Naval Air Engineering Center, Code 5331, Lakehurst, NJ 08733-5000. Eileen Foy (201) 323-2574

Naval Air Propulsion Center, S & T Department (PE3), Trenton, NJ 08628. Albert Martino (609) 896-5713

Naval Weapons Station Earle, Code 70, Colts Neck, NJ 07722. M. Gray (201) 577-2145

NEW MEXICO

Naval Ordnance Missile Test Facility, White Sands, NM 88002-5510. A. F. Schreader (505) 678-2101

PENNSYLVANIA

Naval Air Development Center, Technology Transfer Coordinator, Code 024, Warminster, PA 18974-5000. Jerome S. Bortman (215) 441-2033

VIRGINIA

Marine Corps Development and Education Command, ATTN: Technology Transfer Office, Quantico, VA 22134. R. E. Ouellette (703) 640-3133

DEPARTMENT OF ENERGY

CALIFORNIA

Energy Technology Engineering Center, P.O. Box 1449, Canoga Park, CA 91304. Guy Ervin (818) 700-5532

Laboratory for Energy-Related Health Research, Univ. of California, Davis, CA 95616. Edward A. Rhode (916) 752-1347

Laboratory of Biomedical and Environmental Sciences, Univ. of California, 900 Veteran Ave., Los Angeles, CA 90024. William J. Moffitt (213) 825-9431

Laboratory of Radiobiology and Environmental Health, Univ. of California, School of Medicine, San Francisco, CA 94143. Sheldon Wolff (415) 666-1636

Lawrence Berkeley Laboratory, 50A-4112, Univ. of California, Berkeley, CA 94720. Pepi Ross (415) 486-6502

Lawrence Livermore National Laboratory, Technology Transfer Initiative Program, P.O. Box 808 L-795, Livermore, CA 94550. Gordon Longerbeam (415) 422-6416

Stanford Linear Accelerator Center, Stanford Univ., P.O. Box 4349, Stanford, CA 94309. Herman H. Murphy (415) 926-2130

COLORADO

Rocky Flats Plant, P.O. Box 464, Bldg. 060, Golden, CO 80402-2416. Frederick J. Fraikor (303) 966-2416

Solar Energy Research Institute, 1617 Cole Blvd., Golden, CO 80401. H. Dana Moran (303) 231-7115

IDAHO

Idaho National Engineering Laboratory, INEL-ORTA, P.O. Box 1625, M.S. 3402, Idaho Falls, ID 83415. Richard D. Holman (208) 526-1571

ILLINOIS

Argonne National Laboratory Technology Transfer Center, Bldg. 207, 9700 S. Cass Ave., Argonne, IL 60439. Brian Frost (312) 972-4929

Fermi National Accelerator Laboratory, Fermilab–M.S. 208, P.O. Box 500, Batavia, IL 60510. Richard A. Carrigan (312) 840-3333

New Brunswick Laboratory, 9800 S. Cass Ave., Bldg. 350, Argonne, IL 60439. Carleton D. Bingham (312) 972-2446

INDIANA

Notre Dame Radiation Laboratory, Univ. of Notre Dame, Notre Dame, IN 45556. John Bentley (219) 283-7502

IOWA

Ames Laboratory, Office and Lab. Bldg., Rm. 119, Iowa State Univ., Ames, IA 50011. Daniel E. Williams (515) 294-2635

Massachusetts

William H. Bates Linear Accelerator Center, Mass. Inst. of Tech., P.O. Box 846, Middleton, MA 01949. Ernest J. Moniz (617) 245-6600

Michigan

Michigan State Univ.–DOE Plant Research Laboratory, Michigan State Univ., East Lansing, MI 48824. Gary Watson (517) 353-2270

New Jersey

Princeton Plasma Physics Laboratory, Princeton Univ., P.O. Box 451, Princeton, NJ 08543. Joseph File (609) 243-3009

New Mexico

Inhalation Toxicology Research Institute, Lovelace Biomed. & Envir. Res. Inst., P.O. Box 5890, Albuquerque, NM 87185. Robert Jones (505) 844-2502

Los Alamos National Laboratory, Industrial Initiative Officer, P.O. Box 1663, M899, Los Alamos, NM 87545. Ron Barks (505) 667-3839

Sandia National Laboratories, Technology Transfer Department, Dept. 6010, Albuquerque, NM 87185. Daniel E. Arvizu (505) 846-0387

New York

Brookhaven National Laboratory, Bldg. 475 Office of Res. & Tech., Applications, Upton, Long Island, NY 11973. William Marcuse (516) 282-2103

Environmental Measurement Laboratory, 376 Hudson St., New York, NY 10014-3621. Gail dePlanque (212) 620-3616

Knolls Atomic Power Laboratory, P.O. Box 1072, Schenectady, NY 12301. Gerald Sabien (518) 395-4000

Laboratory for Laser Energetics, Univ. of Rochester, 250 E. River Rd., Rochester, NY 14623. James P. Knauer (716) 275-2074

Ohio

EG & G Mound Applied Technologies, Technology Exchange, P.O. Box 3000, Miamisburg, OH 45343-0987. Japnell D. Braun (513) 865-3829

Pennsylvania

Bettis Atomic Power Laboratory, P.O. Box 1449, West Mifflin, PA 15122. Nick Masterson (412) 476-6111

Pittsburgh Energy Technology Center, P.O. Box 10940, Pittsburgh, PA 15236. Kay Downey (412) 892-6029

South Carolina

Savannah River Ecology Laboratory, P.O. Drawer E, Aiken, SC 29801. Robert I. Nestor (803) 725-2472

Savannah River Laboratory, Aiken, SC 29808. John Corey (803) 725-3020

Tennessee

Oak Ridge Associated Universities, P.O. Box 117, Oak Ridge, TN 37831-0117. Wanda Penland (615) 576-3365

Oak Ridge National Laboratory, P.O. Box 2008, 4500 N. M.S. 257, Oak Ridge, TN 37831-6257. Donald W. Jared (615) 574-4192

Utah

Radiobiology Laboratory, Univ. of Utah, Bldg. 586, Salt Lake, UT 84112. Scott C. Miller (901) 581-7117

Washington

Pacific Northwest Laboratory, P.O. Box 999, Richland, WA 99352. Marv Clement (509) 375-2789

Westinghouse Hanford Laboratory, P.O. Box 1970, M.S. L5-55, Richland, WA 99352. Fred Reich (509) 376-4063

WEST VIRGINIA

Morgantown Energy Technology Center, P.O. Box 880, Collins Ferry Rd., Morgantown, WV 26505. Carol Roberson (304) 291-4308

DEPARTMENT OF HEALTH AND HUMAN SERVICES

ARKANSAS

National Center for Toxicological Research, Food and Drug Administration, HFT-2, Jefferson, AR 72079. Arthur R. Noris (501) 541-4516

MARYLAND

Alcohol, Drug Abuse, and Mental Health Administration, Office of the Administrator 13C-05, 5600 Fishers Ln., Rockville, MD 20857. Thomas R. Vischi (301) 443-4564

Food and Drug Administration, HFY-50, 5600 Fishers Ln., Rockville, MD 20857. Lois Ann Beaver (301) 443-4480

GEORGIA

Centers for Disease Control, 1600 Clifton Rd., NE, Executive Park, Bldg. 24, M.S. E-20, Atlanta, GA 30333. Carl H. Blank (404) 329-1900

OHIO

National Institute of Occupational Safety and Health, 4676 Columbia Pkwy., Cincinnati, OH 45226. Theodore F. Schoenborn (513) 841-4321

DEPARTMENT OF HEALTH AND HUMAN SERVICES; NATIONAL INSTITUTES OF HEALTH

MARYLAND

Frederick Cancer Research Facility, National Cancer Institute, Frederick, MD 21702-1013. Henry J. Hearn (301) 698-1108

National Cancer Institute, Bldg. 32, Rm. 11A23, Bethesda, MD 20892. Elliott Stonehill (301) 496-1148

National Eye Institute, Bldg. 31, 9000 Rockville Pike, Bethesda, MD 20892.

National Heart, Lung, and Blood Institute, Bldg. 31, Rm. 4A21, Bethesda, MD 20205. Larry Blaser (301) 496-4236

National Institute of Allergy and Infectious Diseases, NIH, NIAID, Bldg. 10, Rm. 11C103, Bethesda, MD 20892. Gordon D. Wallace (301) 496-3006

National Institute of Arthritis and Musculoskeletal and Skin Diseases, Westwood Bldg., Rm. 403, 5333 Westbard Ave., Bethesda, MD 20892. Julia Freeman (301) 496-7495

National Institute of Child Health and Human Development, Office of Planning & Eval., NIH, Bldg. 31, Rm. 2A10, Bethesda, MD 20892. James G. Hill (301) 496-1877

National Institute of Dental Research, DPHPB, NIDR WWB, Rm. 522, National Institutes of Health, Bethesda, MD 20892. Alice M. Horowitz (301) 496-2883

National Institute of Diabetes and Digestive and Kidney Diseases, NIH, Bldg. 31, Rm. 9A03, Bethesda, MD 20892. Benjamin Burton (301) 496-4955

National Institute of General Medical Sciences, Westwood Bldg., 5333 Westbard Ave., Bethesda, MD 20892. Dr. Paul Velletri (301) 496-7707

National Institute of Neurological Disorders and Stroke (NINDS), Bldg. 31, Rm. 8A03, NIH, 9000 Rockville Pike, Bethesda, MD 20892. Zekin Shakhashiri (301) 496-9271

National Institute on Aging, Bldg. 31, 9000 Rockville Pike, Bethesda, MD 20892.

National Institutes of Health, Office of Medical Applications of Research, Bldg. 1, Rm. 260, 9000 Rockville Pike, Bethesda, MD 20892. John H. Ferguson (301) 496-5641

NORTH CAROLINA

National Institute of Environmental Health Science, Box 12233, Research Triangle Park, NC 27709. Martin Rodbell (919) 541-3205

DEPARTMENT OF JUSTICE

VIRGINIA

Federal Bureau of Investigation Laboratory, Forensic Science Research and Training Unit, Quantico, VA 22135. Cecil Yates (703) 640-6131

DEPARTMENT OF THE INTERIOR; BUREAU OF MINES

ALABAMA

Tuscaloosa Research Center, P.O. Box L, Univ., Tuscaloosa, AL 35486. Ron Church (205) 758-0491

COLORADO

Denver Research Center, Denver Federal Center, Bldg. 20, Denver, CO 80225. Guy Johnson (303) 236-0747

DISTRICT OF COLUMBIA

Laboratory Technology Transfer, 2401 E. St., NW, Washington, DC 20241. Donald E. Ralston (202) 634-1224

MINNESOTA

Twin Cities Research Center, Bureau of Mines, 5629 Minnehaha Ave., South, Minneapolis, MN 55417. Richard Dick (612) 725-4610

MISSOURI

Rolla Research Center, P.O. Box 280, 1300 Bishop Ave., Rolla, MO 65401. Jim Stephenson (314) 364-3169

NEVADA

Reno Research Center, ATTN: Staff Engineer, U.S. Bureau of Mines, 1605 Evans Ave., Reno, NV 89512-2295. (702) 784-5391

OREGON

Albany Research Center, 1450 Queen Ave., SW, P.O. Box 70, Albany, OR 97321. George J. Dooley (503) 967-5896

PENNSYLVANIA

Pittsburgh Energy Technology Center, P.O. Box 18070, M.S. 59/210, Pittsburgh, PA 15236. Kay Downey (412) 892-4756

UTAH

Salt Lake City Research Center, 729 Arapeen Dr., Salt Lake City, UT 84108-1283. Bill Nissen (801) 524-6113

WASHINGTON

Spokane Research Center, 315 Montgomery Ave., Spokane, WA 99207-2291. Robert Bates (509) 484-1610

DEPARTMENT OF THE INTERIOR; BUREAU OF RECLAMATION

COLORADO

Research and Laboratory Services Division, D-3700C, P.O. Box 25007, Denver Federal Center, Denver, CO 80225. Danny L. King (303) 236-5981

DEPARTMENT OF THE INTERIOR; FISH AND WILDLIFE SERVICE

COLORADO

National Ecology Research Center, 4512 McMurry, Ft. Collins, CO 80522. Frank Dunkle (303) 226-9398

DISTRICT OF COLUMBIA

Cooperative Fish and Wildlife Research Units Center, Cooperative Research Units Center, Washington, DC 20240. Edward T. LaRoe (202) 653-8766

FLORIDA

National Fisheries Research Center–Gainesville, 7920 NW 72st St., Gainesville, FL 32606. James McCann (904) 378-8181

LOUISIANA

National Wetlands Research Center, Fish & Wildlife Service, 1010 Gause Blvd., Slidell, LA 70458. Robert E. Stewart (504) 646-7564

MARYLAND

Patuxent Wildlife Research Center, Laurel, MD 20708. David L. Trauger (301) 498-0211

MICHIGAN

National Fisheries Research Center–Great Lakes, 1451 Green Rd., Ann Arbor, MI 48105. Jon G. Stanley (313) 994-3331

MISSOURI

National Fisheries Contaminants Research Center, Rt. 2, 4200 New Haven Rd., Columbia, MO 65201. Ell-Piret Multer (314) 875-5399

NORTH DAKOTA

Northern Prairie Wildlife Research Center, P.O. Box 2096, Jamestown, ND 58401. Robert B. Oetting (701) 252-5363

WASHINGTON

National Fishery Research Center–Seattle, Bldg. 204, Naval Station, Seattle, WA 98115. Alfred C. Fox (206) 526-6282

WISCONSIN

National Fishery Research Center–LaCrosse, P.O. Box 818, 2630 Fanta Reed Rd., LaCrosse, WI 54602. Fred Meyer (608) 783-6451

National Wildlife Health Center, 6006 Schroeder Rd., Madison, WI 53711. James Kennelly (608) 271-4640

DEPARTMENT OF THE INTERIOR; GEOLOGICAL SURVEY

CALIFORNIA

Geological Survey–Western Region, 345 Middlefield Rd., M.S. 144, Menlo Park, CA 94025. George Gryc (415) 329-4002

COLORADO

Geological Survey–Central Region, M.S. 507, Box 25046, Denver Federal Center, Denver, CO 80225. Lee Aggers (303) 236-5825

VIRGINIA

Geological Survey–National Center, 12201 Sunrise Valley Dr., M.S. 407, Reston, VA 22092. Ethan T. Smith (703) 648-4443

DEPARTMENT OF TRANSPORTATION

COLORADO

Transportation Test Center, P.O. Box 11130, Department of Transportation, Pueblo, CO 81001. David A. Watts (719) 584-0546

CONNECTICUT

Coast Guard Research Development Center, Avery Point, Groton, CT 06340. Samuel F. Powell (203) 441-2602

MASSACHUSETTS

Transportation Systems Center, Code DTS-31, Office of Tech. Sharing, Kendall

Square, Cambridge, MA 02142. R. V. Giangrande (617) 494-2486

New Jersey

Federal Aviation Administration Technology Center, ATTN: ACL-1, Atlantic City Airport, Atlantic City, NJ 08405. A. A. Lupinetti (609) 484-6689

New York

National Maritime Research Center, Kings Point, NY 11024. Walter M. Maclean (516) 482-8200 ext. 571

Ohio

Vehicle Research and Test Center, P.O. Box 37, East Liberty, OH 43319. James E. Hofferberth (513) 666-4511

Oklahoma

Aviation Toxicology Laboratory, FAA Civil Aeromedical Institute, P.O. Box 25082, Oklahoma City, OK 73125. William E. Collins (405) 686-4806

Protection and Survival Laboratory, FAA Aeronautical Center, P.O. Box 25082, Oklahoma City, OK 73125. Richard F. Chandler (405) 686-4851

Virginia

Turner-Fairbank Research Center, Federal Highway Administration, (HRT-1), 6300 Georgetown Pike, McLean, VA 22101. Stanley R. Byington (703) 285-2035

Department of Veterans Affairs

California

Rehabilitation R & D Center, VA Medical Center, 3801 Miranda Ave., Palo Alto, CA 94304. Alvin H. Sacks (415) 858-3991

Georgia

Rehabilitation R & D Center, VA Medical Center, 1670 Clairmont Rd., Decatur, GA 30033. Franklyn K. Coombs (404) 321-5828

Illinois

Rehabilitation R & D Center, Hines VA Hospital, Code: 151L, Box 20, Hines, IL 60141. John Trimble (312) 343-7200 ext. 5780

Maryland

VA Prosthetic R & D Center, Rehabilitation R & D Evaluation Unit, 103 S. Gay St., Baltimore, MD 21202. Mercia C. Decker (301) 962-2333

VA Prosthetic R & D Center, Office of Technology Transfer, 103 S. Gay St., Baltimore, MD 21202. Seldon P. Todd (301) 962-1800

VA Prosthetic R & D Center, 103 S. Gay St., Baltimore, MD 21202. Husher Harris (301) 962-2301

Environmental Protection Agency

California

National Center for Intermedia Transport Research (NCITR), Department of Chemical Engineering, Univ. of California, Los Angeles, CA 90024. Yoram Cohen (213) 825-8766

Western Region Hazardous Substance Research Center, Department of Civil Engineering, Stanford Univ., Stanford, CA 94308. Perry L. McCarty (415) 723-4131

District of Columbia

Office of Toxic Substances, TS 799, 410 M. St., SW, Washington, DC 20460. Bob Jordon (202) 382-3949

FLORIDA

Environmental Research Laboratory (Gulf Breeze), Sabine Island, Gulf Breeze, FL 32561. Raymond G. Wilhour (904) 932-5311

GEORGIA

Environmental Research Laboratory (Athens), US/EPA. College Station Rd., Athens, GA 30613-7799. Rosemarie Russo (405) 332-8800

ILLINOIS

Advanced Environmental Control Technology Center, Department of Civil Engineering, Univ. of Illinois at Urbana Champaign, 205 N. Mathews Ave., Urbana, IL 61801. R. S. Engelbrecht (217) 333-3822

Industrial Waste Elimination Research Center (IWERC), Illinois Institute of Technology, Chicago, IL 60616. James W. Patterson (312) 567-3535

KANSAS

Hazardous Substance Research Center– Region 7/8, Department of Chemical Engineering, Durland Hall, Kansas State Univ., Manhattan, KS 66506. Larry E. Erickson (913) 532-5584

LOUISIANA

Hazardous Waste Research Center, Louisiana State Univ., 3418 CEBA Bldg., Baton Rouge, LA 70803. Louis Thibodeaux (504) 388-6770

MICHIGAN

Grosse Ile Field Station, Gross Ile, MI 48318. William Richardson (313) 675-7704

Hazardous Substance Research Center– Regions 3/5, Department of Civil Engineering, 2340 G. G. Brown, College of Engineering, Univ. of Michigan, Ann Arbor, MI 48109-2125. Walter J. Weber (313) 763-1464

MINNESOTA

Environmental Research Laboratory (Duluth), 6201 Congdon Blvd., Duluth, MN 55804. Nelson Thomas (218) 720-5702

Monticello Field Station, P.O. Box 500, Monticello, MN 55362. Stephen Hedtke (612) 295-5145

NEVADA

Environmental Monitoring Systems Laboratory, P.O. Box 93478, Las Vegas, NV 89193-3478. Robert Snelling (702) 798-2525

NEW JERSEY

Hazardous Substance Research Center– Regions 1/2, New Jersey Institute of Technology, Newark, NJ 07102. Richard S. MaGee (201) 596-3233

Hazardous Waste Engineering Laboratory, Environmental Protection Agency, Edison, NJ 08837. Morris Altschuler (202) 382-7667

Oil and Hazardous Materials Spills Branch, Environmental Protection Agency, Edison, NJ 08837. Gerald Rausa (202) 382-7667

NEW YORK

Ecosystems Research Center, Center for Environmental Research, Corson Hall, Cornell Univ., Ithaca, NY 14953-2701. Leonard H. Weinstein (607) 254-1229

NORTH CAROLINA

Air and Energy Engineering Research Laboratory, USEPA, Research Triangle Park, NC 27711. Frank Princiotta (919) 541-2821

Atmospheric Environment Engineering Research Laboratory, USEPA, Research Triangle Park, NC 27711. Blair Martin (919) 541-2157

Atmospheric Research and Exposure Assessment Laboratory, Office of Research and

Triangle Park, NC 27711. Gary J. Foley (919) 541-2601

Environmental Monitoring Systems Laboratory, Environmental Protection Agency, Research Triangle Park, NC 27711. Morris Altschuler (202) 382-7667

Health Effects Research Laboratory, Office of Research and Development, Office of Health Research, Research Triangle Park, NC 27711. Lawrence Reiter (919) 541-2281

Research Center for Waste Minimization and Management, Department of Chemical Engineering, North Carolina State Univ., Raleigh, NC 27695-7001. Michael R. Overcash (919) 737-2325

OHIO

Environmental Monitoring Systems Laboratory, 26 W. Martin Luther King St., Cincinnati, OH 45268. Thomas A. Clark (513) 569-7301

Hazardous Waste Engineering Research Laboratory, 22 W. St. Clair Rd., Cincinnati, OH 45268. Morris Altschuler (202) 382-7667

Newtown Field Station, Newtown, OH 45244. James Lazorchak (513) 533-8114

Risk Reduction Engineering Laboratory, 26 W. Martin Luther King St., Cincinnati, OH 45268. Louis W. Lefke (513) 569-7953

Water Engineering Research Laboratory, 26 W. Martin Luther King St., Cincinnati, OH 45268. Louis W. Lefke (513) 569-7953

OKLAHOMA

Robert S. Kerr Environmental Research Laboratory (Ada), P.O. Box 1198, Ada, OK 74820. Clinton W. Hall (405) 332-8800

OREGON

Environmental Research Laboratory (Corvallis), USEPA/ERL, 200 Southwest 35th St., Corvallis, OR 97333. Thomas Murphy (503) 757-4601

PENNSYLVANIA

Center for Environmental Epidemiology (CEE), Graduate School of Public Health, Univ. of Pittsburgh, 130 DeSoto St., Pittsburgh, PA 15261. Bruce W. Case (412) 624-1559

RHODE ISLAND

Environmental Research Laboratory (Narragansett), Narragansett, RI 02882. Norbert Jaworski (401) 789-3001

Marine Sciences Research Center (MSRC), Graduate School of Oceanography, Univ. of Rhode Island, Kingston, RI 02881-1197. Michael E. Q. Pilson (401) 792-6104

TEXAS

National Center for Groundwater Research, Dept. of Environmental Science and Engineering, Rice Univ., P.O. Box 1892, Houston, TX 77251. C. H. Ward (713) 527-4086

VIRGINIA

Environmental Photographic Interpretation Center, Bldg. 166, Vint Hill Farm Station (VHFS), Warrenton, VA 22186. (703) 349-8970

WEST VIRGINIA

National Small Flows Clearinghouse, West Virginia Univ., P.O. Box 6064, Morgantown, WV 26506. Stephen Dix (800) 624-8301

FEDERAL EMERGENCY MANAGEMENT AGENCY

OHIO

Disaster Research Center, Ohio State Univ., 128 Derby Hall, Oval Mall, Columbus, OH 43210. Henry Quarantelli (614) 422-5916

NATIONAL AERONAUTICS AND SPACE ADMINISTRATION

ALABAMA

George C. Marshall Space Flight Center, NASA, Mail Code AT01, Huntsville, AL 35812. Ismail Akbay (205) 554-0962

CALIFORNIA

Ames Research Center, Mail Code 223-3, Moffett Field, CA 94035. Geoffrey Lee (415) 604-6406

Jet Propulsion Laboratory, 4800 Oak Grove Dr., M.S. 180-801, Pasadena, CA 91109. Gordon S. Chapman (818) 354-8300

FLORIDA

John F. Kennedy Space Center, NASA, Mail Code: PT-PMO-A, Kennedy Space Center, FL 32899. Thomas M. Hammond (305) 867-3017

MARYLAND

Goddard Space Flight Center, Technology Utilization/Transfer, Mail Code 702, Greenbelt, MD 20771. Donald S. Friedman (301) 286-6242

MISSISSIPPI

John C. Stennis Space Center, Bldg. 1103, Mail Code HA-00, Stennis Space Center, MS 39529. Rick Galle (601) 688-1929

OHIO

Lewis Research Center, LE RC, M.S. 7-3, NASA, 21000 Brookpark Rd., Cleveland, OH 44135. Anthony F. Ratajczak (216) 433-5567

TEXAS

Lyndon B. Johnson Space Center, JSC, Mail Code IC4, NASA, Houston, TX 77058. Dean C. Glenn (713) 483-3809

VIRGINIA

Langley Research Center, M.S. 139A, Hampton, VA 23665-5225. Joe Mathis (804) 864-2484

NATIONAL SCIENCE FOUNDATION

ARIZONA

National Optical Astronomy Observatories, 950 N. Cherry Ave., Tucson, AZ 85719. Sidney C. Wolff (602) 325-9361

COLORADO

National Center for Atmospheric Research, P.O. Box 3000, 1850 Table Mesa Dr., Boulder, CO 80307. Robert Serafin Anderson (303) 497-1108

NEW MEXICO

National Radio Astronomy Observatory, P.O. Box O, Socorro, NM 87801. Ina W. Cole (505) 835-7309

National Solar Observatory, Sunspot, NM 88349. Frank A. Hegwer (505) 434-1390

NEW YORK

National Astronomy and Ionosphere Center, Cornell Univ., Space Sciences Bldg., Ithaca, NY 14853. Tor Hagfors (607) 255-0606

VIRGINIA

National Radio Astronomy Observatory, Edgemont Rd., Charlottesville, VA 22903-2575. Paul A. Vanden Bout (804) 296-0241

SMITHSONIAN INSTITUTION

DISTRICT OF COLUMBIA

National Museum of Natural History, NHB 401, 10th & Constitution Ave., NW, Washington, DC 20560. Stanwyn Shetler (202) 357-2661

National Zoological Park and Conservation Center, National Zoological Park, Washington, DC 20008. Devro G. Kleinman (202) 673-4705

FLORIDA

Smithsonian Tropical Research Institution, Balboa, Canal Zone, APO, Miami, FL 34002. Frank Morris

MARYLAND

Radiation Biology Laboratory, RBL, 12441 Parkland Dr., Rockville, MD 20852. William H. Klein (301) 443-2329

Smithsonian Environmental Research, P.O. Box 28, Edgewater, MD 20137. David Correll (301) 798-4424

MASSACHUSETTS

Smithsonian Astrophysical Observatory, SAO, 60 Garden St., Cambridge, MA 02138. Irwin Shapiro (617) 495-7100

TENNESSEE VALLEY AUTHORITY

ALABAMA

Biomass Branch Laboratory, Office of Agricultural and Chemical Development, 435 Chemical Engineering Bldg., Muscle Shoals, AL 35660. John F. Phillips (205) 386-3065

Western Area Radiological Laboratory, Office of Nuclear Power, Western Area Radiological Laboratory, Muscle Shoals, AL 35660. William L. Raines (205) 386-3758

TENNESSEE

Aquatic Research Laboratory, Department of Aquatic Biology, Haney Bldg. 25 270 C, Chattanooga, TN 37402-2801. Donald C. Wade (202) 386-2068

Engineering Laboratory (Powerplants and Reservoirs), River Basin Operations, Engineering Lab., P.O. Drawer E, Norris, TN 37828. E. Ely Driver (615) 632-1903

Fisheries Laboratory, Office of Natural Resources and Economic Development, Forestry Bldg., Norris, TN 37828. Robert Wallus (615) 632-1797

Tennessee Valley Authority, 400 W. Summit Hill Dr., 2D46 OCH, Knoxville, TN 37902. H. Brown Wright (615) 632-6435

TVA Fossil Plants Laboratories, Office of Power, 2N 66A Blue Ridge Place, Chattanooga, TN 37402-2801. Robert G. Axley (615) 751-2745

INFORMATION CENTERS

Central Industrial Applications Center (CIAC), NASA, Rural Enterprises, Inc., P.O. Box 1335, 10 Waldon Dr., Durant, OK 74702. Dickie Deel (405) 924-5094

Computer Software and Management Information Center (COSMIC), NASA, 382 E. Broad St., Univ. of Georgia, Athens, GA 30602. John A. Gibson (404) 542-3265

Data & Analysis Center for Software (DACS), Rome Air Development Center, Department of the Air Force, ATTN: RADC/COED, Griffiss AFB, NY 13441-5700. Thomas R. Robbins (315) 330-3395

Federal Computer Products Center, National Technical Information Service, Department of Commerce, 5285 Port Royal Rd., Springfield, VA 22161. (703) 487-4763

Food and Nutrition Information Center, Department of Agriculture, National Agricultural Library, 10301 Baltimore Blvd., Beltsville, MD 20705. (301) 344-3719

Government-Industry Data Exchange Program (GIDEP), Department of the Navy, Naval Fleet Analysis Center, Corona, CA 91720. James Richards (714) 736-4677

Manufacturing Technology Information Analysis Center (MTIAC), Department of Defense, 10 West 35th St., Chicago, IL 60616. Michal Stevens (312) 567-4733

Metal Matrix Composites Information Analysis Center (MMCIAC), Department of Defense, Kaman-TEMPO, P.O. Drawer QQ, Santa Barbara, CA 93102. Louis A. Gonzalez (805) 963-6482

Metals and Ceramics Information Center (MCIC), Department of Defense, Battelle-Columbus Division, 505 King Ave., Columbus, OH 43201-2693. (614) 424-5000

NASA/UK Technology Applications Center (NASA/UK-TAC), Univ. of Kentucky, 109 Kinkead Hall, Lexington, KY 40506-0057. William R. Strong (606) 257-6322

National Appropriate Technology Assistance Service (NATAS), Department of Energy, P.O. Box 2525, Butte, MT 59702-2525. (800) 428-2525

National Center for Biotechnology Information, National Library of Medicine, National Institutes of Health, Department of Health and Human Services, 8600 Rockville Pike, Bethesda, MD 20894. (301) 496-2475

National Energy Software Center, Argonne National Laboratory, Department of Energy, 9700 S. Cass Ave., Argonne, IL 60439. (708) 972-7250

National Technical Information Service (NTIS), Department of Commerce, 5285 Port Royal Rd., Springfield, VA 22161. (703) 487-4650

Plastics Technical Evaluation Center (PLASTEC), Army Armament Research, Development and Engineering Center, Department of the Army, ARDEC, SMCAR-AET-O, Picatinny Arsenal, NJ 07806-5000. Suseela Chandrasekar (201) 724-4222

Standard Reference Data, National Institute of Standards and Technology, Department of Commerce, Gaithersburg, MD 20899. Dr. Malcolm W. Chase (301) 975-3692

Northeast Region RTTC, The Center for Technology Commercialization, 100 North Dr., Massachusetts Technology Park, Westborough, MA 01581. Dr. William Gasko, NASA Program Director, James P. Dunn, Assistant (508) 836-4776 (508) 870-0042

Southeast Region RTTC, Southern Technology Applications Center (STAC), Univ. of Florida, 1 Progress Blvd., Box 24, Alachua, FL 32615. J. Ronald Thornton, Director (904) 462-2913 (800) 354-4832 (FL only) (800) 872-7477 (Toll Free U.S.)

Mid-Atlantic Region RTTC, Univ. of Pittsburgh, 823 William Pitt Union, Pittsburgh, PA 15260. Lani Hummel, Director (412) 648-7000

Midwest Region RTTC, Battelle Memorial Institute, Columbus Operations, 505 King Ave., Columbus, OH 43021-2693. Gloria Miller, Contracting Officer (614) 424-7092

Midcontinent Region RTTC, Texas Engineering Experiment Station, Texas A & M Univ. System, 308 Wisenbaker Engineering Research Center, College Station, TX 77843-3124. Gary Sera, Director (409) 845-0538.

Far West Region RTTC, Univ. of Southern California, Research Annex, 3716 S. Hope St., Los Angeles, CA 90007-4334. Robert Stark, Director (213) 743-6132 (800) 642-2872 (CA only) (800) 872-7477 (Toll Free U.S.)

Small Business Innovation Research (SBIR) Contacts

Department of Commerce

DOC SBIR Program Manager, Suitland Professional Center, SPC, Rm. 307, Suitland, MD 20233. Edward V. Tiernan (301) 763-4240

Director, Office of Small and Disadvantaged Business Utilization, U.S. Department of Commerce, 14th and Constitution Ave., NW, HCHB, Rm. 6411, Washington, DC 20230. James P. Maruca (202) 377-1472

Department of Defense

SBIR Program Manager, OSD/SADBU, U.S. Department of Defense, The Pentagon–Rm. 2A340, Washington, DC 20301-3061. Robert Wrenn (202) 697-9383; SDID SBIR Program (800) 937-3150

Department of Education

SBIR Program Coordinator, U.S. Department of Education, Rm. 602F, 555 New Jersey Ave., NW, Washington, DC 20208. John Christensen (202) 219-2065

Department of Energy

DOE Small Business Innovation Research Program, SBIR Program Manager, ER-16, U.S. Department of Energy, Washington, DC 20585. (301) 353-5707

National Aeronautics and Space Administration

Program Director, SBIR Office–Code CR, NASA, 1225 Jefferson Davis Hwy., #1304, Arlington, VA 22202. Harry Johnson (703) 271-5659

National Science Foundation

SBIR Program Managers, National Science Foundation, 1800 St. NW, Washington, DC 20550. Roland Tibbetts, Ritchie Coryell, Darryl G. Gorman (202) 357-7527

Nuclear Regulatory Commission

SBIR Program Representative, Program Management, Policy Development and Analysis Staff, U.S. Nuclear Regulatory Commission, Washington, DC 20555. Marianne M. Riggs (301) 492-3625

Federal Laboratory Consortium Contacts

FLC Locator, DelaBarre & Associates, Inc., 1007 5th Ave., Ste. 810, San Diego, CA 92101. Dr. Andrew Cowan (206) 683-1005

FLC Administrator, Delabarre & Associates, Inc., P.O. Box 545, 224 W. Washington, Suite 3, Sequim, WA 98382-0545. (206) 683-1005

FLC Chairman, Pacific Northwest Laboratory, P.O. Box 999, M.S. K1-34, Richland, WA 99352. Dr. Loren C. Schmid (509) 375-2559

FLC Vice-Chairman, Naval Underwater Systems Center, Code 105–Bldg. 80T, New London, CT 06320. Margaret McNamara (203) 440-4590

FLC REGIONAL COORDINATORS

NORTHEAST REGION

DOT-Federal Aviation Administration Technical Center, Atlantic City, NJ 08405. Al Lupinetti (609) 484-6689

MID-ATLANTIC REGION

DOD-Naval Research Laboratory, Washington, DC 20375-5000. Richard Rein (202) 767-3744

SOUTHEAST REGION

Tennessee Valley Authority, Knoxville, TN 37902. H. Brown Wright (615) 632-6435

MIDWEST REGION

DOE-Argonne National Laboratory, Argonne, IL 60439. Paul Betten (708) 252-5361

MIDCONTINENT REGION

NOAH Environmental Research Lab, 325 Broadway, Boulder, CO 80303. Dr. D. L. Bob Trotter (303) 497-6914

FAR WEST REGION

DOD-Naval Ocean Systems Center, San Diego, CA 92152. Diana Jackson (619) 553-2101

Washington, DC, FLC Representative, 1550 M St., NW, 11th Floor, Washington, DC 20005. Dr. Beverly Berger (202) 331-4220

OTHER USEFUL CONTACTS

Small Business Development Center, President, National Association of Small Business Development Centers, Ames, IA 50010. Ronald A. Manning (515) 292-6351

The National Environmental Technology Applications Corporation (NETAC), Univ. of Pittsburgh Applied Research Center, 615 William Pitt Way, Pittsburgh, PA 15238. (412) 826-5511

Midwest Research Institute Ventures (MRIV), President, 425 Volker Blvd., Kansas City, MO 64110. Robert Muir (816) 753-7600

Argonne, Univ. of Chicago, President and CEO, 1101 East 58th St., Walker Hall, 213B, Chicago, IL 60637. Steven Lazarus (312) 702-1692

Government Microcircuits Applications Conference (GOMAC), Electronic Technology and Devices Laboratory, SLCET-DT, Ft. Monmouth, NJ 07703. Herb Mette (908) 544-3644

National Technology Transfer Center (NTTC), Executive Director, Wheeling Jesuit College, 316 Washington Ave., Wheeling, WV 26003. Lee W. Rivers (304) 243-2455

U.S. Army Civil Engineering Research Laboratory (CERL), 1003 W. Nevada St., Urbana, IL 61801. Robert S. Gorham (217) 333-1369, (800) USA-CERL, Outside Illinois: (800) 252-7122

Strategic Defense Initiative Organization, Office of Technology Applications, Director, Technology Applications, Department of Defense, Washington, DC 20301-7100. Daniel H. Heitz, Colonel, USAF (202) 653-1422

APPLYING FOR CERTIFICATION TO RECEIVE MILITARILY CRITICAL TECHNOLOGY DATA: (DD FORM 2345, "MILITARILY CRITICAL TECHNICAL DATA AGREEMENT")

Commander, Defense Logistics Service Center, Attention: DLSC-FBA, Federal Center, Battle Creek, MI 49017-3084. (800) 352-3572

Business Gold IntelliGate^{TM}, Bell Atlantic© Information Service, Systems Engineering and Management Associates (SEMA), Inc., 5111 Leesburg Pike, Suite 808, Falls Church, VA 22041. Gordon Davidson (703) 845-1200

Defense Technical Information Center, Attn: DTIC-DF, IAC Program Manager, Alexandria, VA 22304-6145. (703) 274-6260

Energy-Related Inventions Program, National Institute of Standards and Technology (NIST), Bldg. 411, Rm. A115, Gaithersburg, MD 20899. George P. Lewett (301) 975-5500

Jet Propulsion Laboratory, Manager, Technology Affiliates Program, M.S. 79-21, Pasadena, CA 91109-8099. James A. Rooney, Ph.D. (818) 354-2503, FAX: (818) 354-7282

Research Triangle Institute, Director, Technology Application Team, P.O. Box 12194, Research Triangle Park, NC 27709. Doris Rouse (919) 541-6980, FAX: (919) 541-6221

NASA Center for Aero Space Information (CASI) (formerly Scientific and Technical Information Facility), P.O. Box 8757, Baltimore, MD 21240-0757. Walter M. Heiland (301) 859-5300, ext. 242, 243

NASA Tech Briefs Magazine, NASA CASI, Manager TU Division, P.O. Box 8757, Baltimore, MD 21240-9985. (301) 859-5300, ext. 242, 243

Aerospace Research Applications Center, Indianapolis Center for Advanced Research, 611 N. Capitol Ave., Indianapolis, IN 46204. (317) 262-5003

NERAC, Inc., One Technology Dr., Tolland, CT 06084. Sam DiSavino (203) 872-7000

Appendix II

Technology Transfer Checklist

This checklist is based on the authors' direct experiences and observations of what actions lead to successful technology transfer from the federal government. Use it as a guide to plan your own efforts, preparing yourself with focused objectives and critical insights into the needs of the players on both sides.

The questions in the checklist are organized under the categories of why, what, where, when, how, and who. This simple arrangement allows for easier recall so that a mental form of the checklist can be easily carried into both technical discussions and agreement negotiations. Space is provided at the end of each list to add your own questions.

I. **Why?**
   Don't neglect to ask these questions. The answers cannot be assumed to be self-evident and, without clarity on these points, it will be more difficult to make a long-term commitment to the entire process.

   A. Why are we looking for technology from outside the company?

      1. Do we have technical problems that cannot efficiently be solved internally?

      2. Do we have new product needs and limited development time or resources?

      3. Can we enter a new market faster this way?

      4. What is the competition doing?

   B. Why are we looking to the federal labs for technology?

      1. Is there a technology under development at a lab that fits our plans?

      2. Is it at the right stage of development?

      3. Does it offer the needed proprietary position?

      4. Are we prepared to exercise the necessary diligence in protecting proprietary data?

      5. Are access costs and reengineering costs going to be within reason?

      6. Is there cofunding, "product maturation," or grants available from the lab/agency?

      7. Is the opportunity for spin-back of a new product to sell to the government important?

      8. Are we in a good position to win a competitive bid for a joint development contract or to justify a sole source relationship?

C. Other

   1.

   2.

   3.

II. **What?**
Knowing ahead of time what you are looking for will make it easier for your government counterpart to get the answers needed to decide if proceeding is worthwhile.

   A. Do we need an answer to a technical problem?

   B. Are our technical objectives clearly defined and well understood?

   C. Do we need a new ready-to-go product?

   D. Are patents important?

   E. Are background patents important/available?

   F. Are we looking for long-range insight into a fundamental discipline?

   G. Do we need access to a unique testing or research facility?

   H. Are we surveying the state-of-the-art and keeping up with what the competition is doing?

   I. Other

   1.

   2.

   3.

III. **Where?**
Is this the lab to work with?

   A. Does this lab have unique facilities that are essential?

   B. Does this lab have a particular scientist or engineer who is essential?

   C. Is this lab the leader in the field and is that essential?

D. Is the location convenient and within the travel/communications budget?

E. Is the atmosphere for technology transfer supportive?

F. Has this lab ever entered a relationship with the private sector similar to the one contemplated?

G. Was it handled smoothly?

H. Other

1.

2.

3.

IV. **When?**
Is the timing right and can the other party deliver?

A. Is this the right time for us to undertake this effort?

B. How much time is available for our staff's involvement?

C. Within what time frame do we need to have usable results?

D. What is the estimate of the government employees as to how long it will take?

E. Does adding a generous margin of error still make the time frame reasonable?

F. Have we been able to agree to dated milestones for completion of crucial steps?

G. What is the expected date of important new product releases from the competition?

H. Other

1.

2.

3.

V. **How?**
Choosing the right mechanism to find and exploit the desired technology.

A. Have we checked all sources of information that are relevant?

B. Should we use the services of a broker?

C. What criteria would be used in the selection of a broker?

D. In our first contact with the lab did we indicate a willingness to make a near-term visit?

E. What type of arrangement do we need/prefer?

   1. Technical information?

   2. Software?

   3. Copyrighted software (from a contractor or assigned to government by contractor)?

   4. Patent license?

   5. CRADA?

   6. Cost-shared development under a contract (open bid or sole source)?

   7. Consulting?

   8. Industrial guest investigator?

   9. SBIR or ATP grant?

   10. Job/partnership offer to a government employee?

F. Are model agreements available?

G. Are intellectual property and data rights (FOIA) concerns taken care of?

H. If a joint effort is contemplated, are the technical objectives mutually desirable?

I. Is there a written, agreed-to, project plan for meeting these technical objectives?

J. Other

   1.

   2.

   3.

VI. **Who?**
   Technology transfer is a person-to-person sport. Don't neglect the human factors.

   A. Have all relevant parties on both sides been informed early in the process?

   B. Who has what responsibilities?

   C. Are the participants clearly aware of their responsibilities?

   D. Who has ultimate oversight and accountability on both sides?

   E. Does everyone have the needed contact information—names, titles, addresses, phone, fax?

   F. Do we sufficiently understand the culture of the other side (e.g., don't offer to buy a government employee lunch)?

   G. What are each person's needs—political, personal, technical, financial?

   H. Are each person's needs being met?

   I. Other

      1.

      2.

      3.

Appendix III

Sample Documents

NATIONAL INSTITUTES OF HEALTH ALCOHOL, DRUG ABUSE AND MENTAL HEALTH ADMINISTRATION COOPERATIVE RESEARCH AND DEVELOPMENT AGREEMENT

This Cooperative Research and Development Agreement, hereinafter referred to as the "CRADA," consists of this Cover Page, an attached Agreement, a Signature Page and various Appendices referenced in the Agreement. This Cover Page serves to identify the Parties to this CRADA:

(1) the following Bureau(s), Institute(s) or Division(s) of the National Institutes of Health and/or the Alcohol, Drug Abuse and Mental Health Administration: _____, hereinafter singly or collectively referred to as the "NIH/ADAMHA," and

(2) _____, which has offices at _____, hereinafter referred to as the "Collaborator."

Although drafted for two Parties, the attached CRADA may also be used for any number. This Cover Page, however, should be modified by repeating block (2) to identify other Parties to the CRADA. All non-NIH/ADAMHA Parties are hereinafter collectively referred to as the "Collaborator." Use of the terms "Collaborator," "Party" and "Parties" should be construed as appropriate for the actual number of CRADA participants.

COOPERATIVE RESEARCH AND DEVELOPMENT AGREEMENT

ARTICLE 1. INTRODUCTION

This Cooperative Research and Development Agreement (CRADA) between NIH/ADAMHA and the Collaborator will be effective when signed by all parties. By signing this CRADA, the Collaborator acknowledges that it has received and read a copy of the Policy Statement on Cooperative Research and Development Agreements and Intellectual Property Licensing which is attached as Appendix A. The research and development project(s) which will be undertaken by each of the Parties in the course of this CRADA are detailed in the Research Plan (RP) attached as Appendix B. The funding and staffing commitments of the Parties are set forth in Appendix C. Any exceptions or

Article 2. DEFINITIONS

As used in this CRADA, the following terms shall have the indicated meanings:

2.1 "Cooperative Research and Development Agreement" or "CRADA" means this Agreement, entered into by NIH/ADAMHA pursuant to the Federal Technology Transfer Act of 1986 and Executive Order 12591 of October 10, 1987.

2.2 "Proprietary Information" means confidential scientific, business or financial information provided that such information:

2.2.1 is not publicly known or available from other sources who are not under a confidentiality obligation to the source of the information;

2.2.2 has not been made available by its owners to others without a confidentiality obligation;

2.2.3 is not already known by or available to the receiving Party without a confidentiality obligation; or

2.2.4 does not relate to potential hazards or cautionary warnings associated with the production, handling or use of the subject matter of the Research Plan of this CRADA.

2.3 "Subject Data" means all recorded information first produced in the performance of this CRADA.

2.4 "Research Results" means all tangible materials other than Subject Data first produced in the performance of this CRADA.

2.5 "Subject Invention" means any invention, conceived or reduced to practice in the performance or research under this CRADA, that may be patentable under 35 U.S.C. § 101 or § 161, protectable under 7 U.S.C. § 2321, or otherwise protectable by other types of U.S. or foreign "Intellectual Property" ("IP") right.

2.6 "Government" means the U.S. Government and any of its agencies.

2.7 "Research Plan" or "RP" means the statement in Appendix B of the respective research and development commitments of the Parties to this CRADA.

2.8 "Principal Investigator" or "PI" means the persons designated respectively by the Parties to this CRADA who will be responsible for the scientific and technical conduct of the RP.

Article 3. COOPERATIVE RESEARCH

3.1 *Research Team.* The Parties agree to establish a joint research and development team (hereinafter referred to as the "Team") comprising at least the Principal

Investigators designated pursuant to Article 3.3 to conduct and monitor the research in accordance with the RP. Although the members of the Team shall be considered as having been delegated to the Team, they shall continue to remain employed by their respective employers under their respective terms of employment.

3.2 *Review of Work*. Periodic conferences shall be held by the Team to review work progress. It is understood that the nature of this cooperative research precludes a guarantee of its completion within the specified period of performance or limits of allocated financial or staffing support. Accordingly, research under this CRADA is to be performed on a best efforts basis.

3.3 *Principal Investigators*. NIH/ADAMHA research work under this CRADA will be performed by the Laboratory identified in the RP, and the NIH/ADAMHA Principal Investigator (PI) designated in the RP will be responsible for the scientific and technical conduct of this project on behalf of NIH/ADAMHA. Also designated in the RP is the Collaborator PI who will be responsible for the scientific and technical conduct of this project on behalf of the Collaborator.

3.4 *Research Plan Change*. The RP may be modified by mutual written consent of the Principal Investigators. Substantial changes in the scope of the RP will be treated as amendments under Article 14.6.

ARTICLE 4. REPORTS

4.1 *Interim Reports*. The Parties shall exchange formal written interim progress reports on a schedule agreed to by the PIs, but at least within six (6) months after this CRADA becomes effective and at least within every six (6) months thereafter. Such reports shall set forth the technical progress made, identifying such problems as may have been encountered and establishing goals and objectives requiring further effort.

4.2 *Final Reports*. The Parties shall exchange final reports of their results within four (4) months after completing the projects described in the RP or after the termination of this CRADA.

ARTICLE 5. FINANCIAL AND STAFFING OBLIGATIONS

5.1 *NIH/ADAMHA and Collaborator Contributions*. The NIH/ADAMHA contribution to the RP in the form of personnel, services and property only is designated in Appendix C. The Collaborator contribution to the RP in the form of personnel, services, property, support for staffing and/or funding is designated in Appendix C. Payment schedules, if applicable, are also indicated in Appendix C.

5.2 *Insufficient and Excess Funds*. NIH/ADAMHA shall not be obligated to perform any of the research specified herein or to take any other action required by this CRADA if the funding is not provided as set forth in Appendix C. NIH/

ADAMHA shall return excess funds to the Collaborator when it sends its final fiscal report pursuant to Article 5.3, except for staffing support pursuant to Article 11.3.

5.3 *Accounting Records.* NIH/ADAMHA shall maintain separate and distinct current accounts, records, and other evidence supporting all its obligations under this CRADA, and shall provide the Collaborator an annual report reflecting the use of the Collaborator's funds and a final such fiscal report at the time that final reports are exchanged pursuant to Article 4.2.

ARTICLE 6. TITLE TO PROPERTY

6.1 *Capital Equipment.* The purchase or use of capital equipment to carry out this CRADA does not affect the ownership rights that would otherwise apply. Equipment purchased by NIH/ADAMHA with funds provided by the Collaborator shall be the property of NIH/ADAMHA. All capital equipment provided under this CRADA by one party for the use of another Party remains the property of the providing Party unless other disposition is mutually agreed upon in writing by the PIs. If title to this equipment remains with the providing Party, that Party is responsible for maintenance of the equipment and the costs of its transportation to and from the site where it will be used.

ARTICLE 7. INTELLECTUAL PROPERTY RIGHTS AND APPLICATIONS

7.1 *Reporting.* The Parties shall promptly report to each other in writing each Subject Invention resulting from the research conducted under this CRADA that is reported to them by their respective employees. Such reports shall be treated in confidence by the receiving Party until such time as a patent or other Intellectual Property (IP) application, as appropriate, claiming that Subject Invention has been filed. Because of the royalty sharing provisions for Government inventors in the Federal Technology Transfer Act of 1986, and in view of Article 8.4 of this CRADA which grants the Government only a research license on inventions made solely by the Collaborator, the Collaborator acknowledges a special duty to report all Subject Inventions to NIH/ADAMHA so that NIH/ADAMHA may determine whether or not inventorship properly includes NIH/ADAMHA investigators.

7.2 *Collaborator Employee Inventions.* The Collaborator may elect to retain IP rights to any Subject Invention made solely by a Collaborator employee. The Collaborator shall notify NIH/ADAMHA promptly upon making this election. If the Collaborator does not elect to retain its IP rights, the Collaborator shall offer to assign these IP rights to the Subject Invention to NIH/ADAMHA pursuant to Article 7.5. If NIH/ADAMHA declines such assignment, the Collaborator may release its IP rights to employee inventors pursuant to Article 7.6.

7.3 *NIH/ADAMHA Employee Inventions.* NIH/ADAMHA on behalf of the U.S. Government may elect to retain IP rights to each Subject Invention made solely by NIH/ADAMHA employees. If NIH/ADAMHA does not elect to retain IP rights, NIH/ADAMHA shall offer to assign these IP rights to such Subject Invention to the Collaborator pursuant to Article 7.5. If the Collaborator declines such assignment, NIH/ADAMHA may release IP rights in such Subject Invention to its employee inventors pursuant to Article 7.6.

7.4 *Joint Inventions*. Each Subject Invention made jointly by NIH/ADAMHA and Collaborator employees shall be jointly owned by NIH/ADAMHA and the Collaborator. The Collaborator may elect to file the joint patent or other IP application(s) thereon and shall notify NIH/ADAMHA promptly upon making this election. If the Collaborator decides to file such applications, it shall do so in a timely manner and at its own expense. If the Collaborator does not elect to file such application(s), NIH/ADAMHA on behalf of the U.S. Government shall have the right to file the joint applications in a timely manner and at its own expense. If either Party decides not to retain its IP rights to a jointly owned Subject Invention, it shall offer to assign such rights to the other Party pursuant to Article 7.5. If the other Party declines such assignment, the offering Party may release its IP rights to employee inventors pursuant to Article 7.6.

7.5 *Filing of Patent Applications*. With respect to Subject Inventions made by the Collaborator as described in Article 7.2 or by NIH/ADAMHA as described in Article 7.3, a Party exercising its right to elect to retain IP rights to a Subject Invention agrees to file patent or other IP applications in a timely manner and at its own expense. The Party may elect not to file a patent or other IP application thereon in any particular country or countries provided it so advises the other Party ninety (90) days prior to the expiration of any applicable filing deadline, priority period or statutory bar date, and hereby agrees to assign its IP right, title and interest in such country or countries to the Subject Invention to the other Party and to cooperate in the preparation and filing of a patent or other IP applications. In any countries in which title to patent or other IP rights is transferred to the Collaborator, the Collaborator agrees that NIH/ADAMHA inventors will share in any royalty distribution that the Collaborator pays to its own inventors.

7.6 *Release to Inventors*. In the event neither of the Parties to this CRADA elects to file a patent or other IP application on a Subject Invention, either or both (if a joint invention) may release their IP rights to their respective employee inventor(s) with a nonexclusive, nontransferrable, royalty-free license being retained by each Party.

7.7 *Patent Expenses*. The expenses attendant to the filing of patent or other IP applications generally shall be paid by the Party filing such application. If an exclusive license to any Subject Invention is granted to the Collaborator, the Collaborator shall reimburse NIH/ADAMHA for the reasonable past and Collaborator-approved ongoing funds expended worldwide for filing, prosecuting and maintaining any applications claiming such exclusively licensed inventions and any patents or other IP grants that may issue on such applications. The Collaborator may waive its exclusive license rights on any application, patent or other IP grant at any time, and incur no subsequent compensation obligation for that application, patent or IP grant.

7.8 *Prosecution of Intellectual Property Applications.* Each Party shall provide the other party with copies of the applications it files on any Subject Invention along with the power to inspect and make copies of all documents retained in the patent or other IP application files by the applicable patent or other IP office. The Parties agree to consult with each other with respect to the prosecution of NIH/ADAMHA Subject Inventions described in Article 7.3 and joint Subject Inventions described in Article 7.4. If the Collaborator elects to file and prosecute IP applications on joint Subject Inventions pursuant to Article 7.4, NIH/ADAMHA will be granted an associate power of attorney (or its equivalent) on such IP applications.

ARTICLE 8. LICENSING

8.1 *Option for Exclusive Commercialization License.* With respect to Government IP rights to any Subject Invention not made solely by the Collaborator's employees for which a patent or other IP application is filed, NIH/ADAMHA hereby grants to the Collaborator an option to negotiate, in good faith, the terms of an exclusive or nonexclusive commercialization license that fairly reflect the relative contributions of the Parties to the invention and the CRADA, the risks incurred by the Collaborator and the costs of subsequent research and development needed to bring the invention to the marketplace. The license will specify the licensed fields of use, breadth of exclusivity and royalties. Royalty rates will be based on product sales and the rates conventionally granted in the field identified in the RP for inventions with reasonably similar commercial potential. Royalty rates generally will not exceed a rate within the range of 5% to 8% for exclusive commercialization licenses. Contingent royalty schemes based on, e.g., patent issuance or nonissuance, and provisions treating the stacking of royalties or packaging of other licensed inventions developed under this CRADA may be provided. Exclusive licensees will be expected to reimburse NIH/ADAMHA for IP expenses related to each licensed intellectual property, and may be permitted to offset such reimbursement against future product royalties.

8.2 *Exercise of License Option.* The option of Article 8.1 must be exercised by written notice mailed within three (3) months after the patent or other IP application is filed to the NIH Office of Invention Development, Building 31, Room B1C38, National Institutes of Health, Bethesda, MD 20892. Exercise of this option by the Collaborator initiates a negotiation period that expires nine (9) months after the patent or other IP application filing date. If the last proposal by the Collaborator has not been responded to in writing by NIH/ADAMHA within this nine (9) month period, the negotiation period shall be extended to expire one (1) month after NIH/ADAMHA so responds, during which month the Collaborator may accept in writing the final license proposal of NIH/ADAMHA. After that time, NIH/ADAMHA will be free to license such IP rights to others.

8.3 *Pricing.* NIH/ADAMHA have a concern that there be a reasonable relationship between the pricing of a licensed product, the public investment in that product, and the health and safety needs of the public. Accordingly, exclusive commercialization licenses granted for NIH/ADAMHA intellectual property rights may require that this relationship be supported by reasonable evidence.

8.4 *Government Intellectual Property Rights.* For inventions developed wholly by NIH/ADAMHA investigators or jointly with a Collaborator under this CRADA, NIH/ADAMHA are required by the Federal Technology Transfer Act of 1986, 15 U.S.C. at § 3710a(b)(2), to retain at least a nonexclusive, irrevocable, paid-up license to practice the invention or to have the invention practiced throughout the world by or on behalf of the U.S. Government. For inventions developed wholly by the Collaborator under this CRADA, the Collaborator agrees to grant a research license as described in Article 8.5 to the Government.

8.5 *Research Licenses.* NIH/ADAMHA reserve the right under any IP license granted to the Collaborator under this CRADA to grant nonexclusive licenses to third parties to make and to use the licensed invention for purposes of research involving the invention itself, and not for purposes of commercial manufacture or in lieu of purchase as a commercial product for use in other research. NIH/ADAMHA intend to consult with their exclusive commercialization licensee(s) before granting research licenses to commercial entities.

8.6 *Joint Inventions Not Exclusively Licensed.* In the event that the Collaborator does not acquire an exclusive commercialization license to IP rights in joint Subject Inventions described in Article 7.4, then each Party shall have the right to use the joint Subject Invention and to license its use to others. The Parties may agree to a joint licensing approach for such IP rights.

ARTICLE 9.   PROPRIETARY RIGHTS AND PUBLICATION

9.1 *Right of Access.* NIH/ADAMHA and the Collaborator agree to exchange all Subject Data and Research Results produced in the course of research under this CRADA, whether developed solely by NIH/ADAMHA, jointly with the Collaborator, or solely by the Collaborator. Tangible research products developed under a CRADA will be shared equally by the Parties to the CRADA unless other disposition is agreed to by the PIs. All Parties to this CRADA will be free to utilize Subject Data and Research Results for their own purposes, consistent with their obligations under this CRADA.

9.2 *Ownership of Subject Data and Research Results.* Subject to the sharing requirements of Article 9.1, the producing Party will retain ownership of and title to all

Subject Inventions, all Subject Data and all Research Results produced solely by their investigators. Jointly developed Subject Inventions, Subject Data and Research Results will be jointly owned. However, except as may be afforded through IP rights that require public disclosure of the protected subject matter (e.g., patents), NIH/ADAMHA do not have statutory authority to license (or agree with the Collaborator to limit dissemination) of Subject Data or Research Results developed solely by NIH/ADAMHA investigators or jointly with the Collaborator. Accordingly, NIH/ADAMHA will not agree to exclude others from utilizing or commercializing such Subject Data or Research Results.

9.3 *Proprietary and Confidential Information.* Each Party agrees to limit its disclosure of Proprietary Information to the amount necessary to carry out the Research Plan of this CRADA, and shall place a confidentiality notice on all such information. Research materials required for the RP may also be designated as Proprietary Information. Each Party receiving Proprietary Information agrees that any information so designated shall be used by it only for the purposes described in the attached Research Plan. Any Party may object to the designation of information as Proprietary Information by another Party and may decline to accept such information. Data and research products developed solely by the Collaborator may be designated as Proprietary Information when they are wholly separable from the data and research products developed jointly with NIH/ADAMHA investigators, and advance designation of such data and product categories is set forth in the RP. The exchange of confidential information, e.g., patient data, should be similarly limited and treated. Unless disclosure is otherwise mutually agreed upon, all Parties to this CRADA agree to keep CRADA Subject Data and Research Results confidential, to the extent permitted by law, until they are published, or corresponding patent or other IP application(s) have been filed.

9.4 *Protection of Proprietary Information.* Proprietary Information shall not be disclosed, copied, reproduced or otherwise made available to any other person or entity without the consent of the owning Party except as required under court order or the Freedom of Information Act (5 U.S.C. § 552). Each Party agrees to use its best efforts to maintain the confidentiality of Proprietary Information. Each Party agrees that another Party is not liable for the disclosure of Proprietary Information which, after notice to and consultation with the concerned Party, another Party in possession of the Proprietary Information determines may not lawfully be withheld, provided the concerned Party has been given an opportunity to obtain a court order to enjoin disclosure.

9.5 *Duration of Confidentiality Obligation.* The obligation to maintain the confidentiality of Proprietary Information shall expire at the earlier of the date when the information is no longer Proprietary Information as defined in Article 2.2 or three (3) years after the expiration or termination date of this CRADA. The Collaborator may request an extension to this term when necessary to protect Proprietary Information relating to products not yet commercialized.

9.6 *Publication.* The Parties are encouraged to make publicly available the results of their research. Before either Party submits a paper or abstract for publication or otherwise intends to publicly disclose information about a Subject Invention, Subject Data or Research Results, the other Party shall be provided thirty (30) days to review the proposed publication or disclosure to assure that Proprietary Information is protected. The publication or other disclosure shall be delayed for up to thirty (30) additional days upon written request by any Party as necessary to preserve U.S. or foreign patent or other IP rights.

ARTICLE 10. REPRESENTATIONS AND WARRANTIES

10.1 *Representations and Warranties of NIH/ADAMHA.* NIH/ADAMHA hereby represents and warrants to the Collaborator that the Official signing this CRADA has authority to do so.

10.2 *Representations and Warranties of the Collaborator.* The Collaborator hereby represents and warrants to NIH/ADAMHA that the Collaborator has the requisite power and authority to enter into this CRADA and to perform according to its terms, and that the Collaborator's Official signing this CRADA has authority to do so. The Collaborator further represents that it is financially able to satisfy any funding commitments made in Appendix C.

ARTICLE 11. TERMINATION

11.1 *Termination by Mutual Consent.* NIH/ADAMHA and the Collaborator may terminate this CRADA, or portions thereof, at any time by mutual written consent. In such event the Parties shall specify the disposition of all property, inventions, patent or other IP applications and other results of work accomplished or in progress, arising from or performed under this CRADA.

11.2 *Unilateral Termination.* Either NIH/ADAMHA or the Collaborator may unilaterally terminate this entire CRADA at any time by giving written notice at least thirty (30) days prior to the desired termination date, and any rights accrued in property, patents or other IP shall be disposed of as in 11.1.

11.3 *Staffing.* If this CRADA is mutually or unilaterally terminated prior to its expiration, funds will nevertheless remain available to NIH/ADAMHA for continuing any staffing commitment made by the Collaborator pursuant to Article 5.1 above and Appendix C, if applicable, for a period of six (6) months after such termination. If there are insufficient funds to cover this expense, the Collaborator agrees to pay the difference.

11.4 *New Commitments.* No Party shall make new commitments related to this CRADA after a mutual or unilateral termination and shall, to the extent feasible, cancel all outstanding commitments and contracts by the termination date.

11.5 *Termination Costs.* Concurrently with the exchange of final reports pursuant to Articles 4.2 and 5.3, NIH/ADAMHA shall submit to the Collaborator for payment a statement of all costs incurred prior to the date of termination and for all reasonable termination costs including the cost of returning Collaborator property or removal of abandoned property.

ARTICLE 12. DISPUTES

12.1 *Settlement.* Any dispute arising under this CRADA which is not disposed of by agreement of the Principal Investigators shall be submitted jointly to the signatories of this CRADA. If the signatories are unable to resolve the dispute jointly within thirty (30) days after notification thereof, the Assistant Secretary of Health (or his/her designee) shall propose a resolution. Nothing in this section shall prevent any Party from pursuing any and all administrative and/or judicial remedies which may be available.

12.2 *Continuation of Work.* Pending the resolution of any dispute or claim pursuant to this Article, the Parties agree that performance of all obligations shall be pursued diligently in accordance with the direction of the NIH/ADAMHA signatory.

ARTICLE 13. LIABILITY

13.1 *Property.* The U.S. Government shall not be responsible for damages to any property of the Collaborator provided to it or acquired by it pursuant to this CRADA.

13.2 *No Warranties.* Except as specifically stated in Article 10, the Parties make no express or implied warranty as to any matter whatsoever, including the conditions of the research or any invention or product, whether tangible or intangible, made, or developed under this CRADA, or the ownership, merchantability, or fitness for a particular purpose of the research or any invention or product.

13.3 *Indemnification.* The Collaborator agrees to hold the U.S. Government harmless and to indemnify the Government for all liabilities, demands, damages, expenses and losses arising out of the use by the Collaborator for any purpose of the Subject Data, Research Results and/or Subject Inventions produced in whole or part by NIH/ADAMHA employees under this CRADA, unless due to the negligence of NIH/ADAMHA, its employees or agents. The Collaborator shall be liable for any claims or damages it incurs in connection with this CRADA. NIH/ADAMHA have no authority to indemnify the Collaborator.

13.4 *Force Majeure.* Neither Party shall be liable for any unforeseeable event beyond its reasonable control not caused by the fault or negligence of such Party, which causes such Party to be unable to perform its obligations under this CRADA, and which it has been unable to overcome by the exercise of due diligence. In the event of occurrence of such a *force majeure* event, the Party unable to perform shall promptly notify the other Party. It shall further use its best efforts to resume performance as quickly as possible and shall suspend performance only for such period of time as is necessary as a result of the *force majeure* event.

Article 14. Miscellaneous

14.1 *Governing Law.* The construction, validity, performance and effect of this CRADA shall be governed by Federal law, as applied by the Federal Courts in the District of Columbia. Federal law and regulations will preempt any conflicting or inconsistent provisions in this CRADA.

14.2 *Entire Agreement.* This CRADA constitutes the entire agreement between the Parties concerning the subject matter of this CRADA and supersedes any prior understanding or written or oral agreement.

14.3 *Headings.* Titles and headings of the sections and subsections of this CRADA are for the convenience of reference only, do not form a part of this CRADA and shall in no way affect its interpretation.

14.4 *Waivers.* None of the provisions of this CRADA shall be considered waived by any Party unless such waiver is given in writing to the other Party. The failure of a Party to insist upon strict performance of any of the terms and conditions hereof, or failure or delay to exercise any rights provided herein or by law, shall not be deemed a waiver of any rights of any Party.

14.5 *Severability.* The illegality or invalidity of any provisions of this CRADA shall not impair, affect or invalidate the other provisions of this CRADA.

14.6 *Amendments.* If either Party desires a modification to this CRADA, the Parties shall, upon reasonable notice of the proposed modification or extension by the Party desiring the change, confer in good faith to determine the desirability of such modification or extension. Such modification shall not be effective until a written amendment is signed by the signatories to this CRADA or by their representatives duly authorized to execute such amendment.

14.7 *Assignment.* Neither this CRADA nor any rights or obligations of any Party hereunder shall be assigned or otherwise transferred by either Party without the prior written consent of the other Party.

14.8 *Notices.* All notices pertaining to or required by this CRADA shall be in writing and shall be signed by an authorized representative and shall be delivered by hand or sent by certified mail, return receipt requested, with postage prepaid, to the addresses indicated on the signature page for each Party. Notices regarding the exercise of license options shall be made pursuant to Article 8.2. Any Party may change such address by notice given to the other Party in the manner set forth above.

14.9 *Independent Contractors.* The relationship of the Parties to this CRADA is that of independent contractors and not as agents of each other or as joint venturers or partners. Each Party shall maintain sole and exclusive control over its personnel and operations. Collaborator employees who will be working at NIH/ADAMHA facilities may be asked to sign a Guest Researcher or Special Volunteer Agreement appropriately modified in view of the terms of this CRADA.

14.10 *Use of Name or Endorsements.* By entering into this CRADA, NIH/ADAMHA does not directly or indirectly endorse any product or service provided, or to be provided, whether directly or indirectly related to either this CRADA or to any patent or other IP license or agreement which implements this CRADA by its successors, assignees or licensees. The Collaborator shall not in any way state or imply that this CRADA is an endorsement of any such product or service by the U.S. Government or any of its organizational units or employees.

14.11 *Exceptions to This CRADA.* Any exceptions or modifications to this CRADA that are agreed to by the Parties prior to their execution of this CRADA are set forth in Appendix D.

14.12 *Reasonable Consent.* Whenever a Party's consent or permission is required under this CRADA, such consent or permission shall not be unreasonably withheld.

Article 15. DURATION OF AGREEMENT

15.1 *Duration.* It is mutually recognized that the duration of this project cannot be rigidly defined in advance, and that the contemplated time periods for various phases of the RP are only good faith guidelines subject to adjustment by mutual agreement to fit circumstances as the RP proceeds. In no case will the term of this CRADA extend beyond the term indicated in the RP unless it is revised in accordance with Article 14.6.

15.2 *Survivability.* The provisions of Articles 4.2, 5.2, 5.3, 6.1, Articles 7–9, 11.3, 11.5, 12.1, 13.3 and 14.10 shall survive the termination of this CRADA.

CRADA Signature Page

FOR NIH/ADAMHA:

_____   _____
                                                Date

_____

Mailing Address for Notices:

_____
_____
_____
_____

FOR THE COLLABORATOR:

_____   _____
                                        Date
_____

Mailing Address for Notices:
_____
_____
_____
_____
_____

[Include additional signature and address blocks as necessary for all Parties to this CRADA.]

APPENDIX A
NIH/ADAMHA POLICY STATEMENT ON COOPERATIVE RESEARCH AND DEVELOPMENT AGREEMENTS AND INTELLECTUAL PROPERTY LICENSING

NATIONAL INSTITUTES OF HEALTH, ALCOHOL, DRUG ABUSE AND MENTAL HEALTH ADMINISTRATION POLICY STATEMENT ON COOPERATIVE RESEARCH AND DEVELOPMENT AGREEMENTS AND INTELLECTUAL PROPERTY LICENSING

This Statement sets forth the policies of the National Institutes of Health (NIH) and the Alcohol, Drug Abuse and Mental Health Administration (ADAMHA) on various aspects of cooperative research and intellectual property licensing. These policies apply to the negotiation of NIH/ADAMHA Cooperative Research and Development Agreements (CRADAs). License agreements for intellectual property rights to inventions developed under a CRADA or through the NIH/ADAMHA intramural research programs, whether negotiated by NIH/ADAMHA or the National Technical Information Service on their behalf, will also incorporate these policies. This Statement may be revised from time to time as NIH and ADAMHA consider appropriate.*

To implement the Federal Technology Transfer Act of 1986 (FTTA, 15 U.S.C. at § 3710), Executive Order 12591 of April 10, 1987 orders Federal laboratories to assist universities and the private sector in broadening our national technology base by moving new knowledge from the research laboratory into the development of new products and processes. Although Federal patent law (35 U.S.C. at §§ 200-212) authorizes the licensing of Government-owned patent rights, the FTTA seeks to facilitate technological collaboration at an earlier stage. Thus, the FTTA authorizes Federal laboratories to enter into CRADAs, and to agree to grant intellectual property rights in advance to collaborators for inventions made in whole or part by Federal employees under the CRADA. Besides assisting in the transfer of commercially useful technologies from Federal laboratories to the marketplace, CRADAs make outside resources more accessible to Federal laboratories.

NIH and ADAMHA, agencies of the Public Health Service (PHS) within the Department of Health and Human Services (DHHS), are among the world's preeminent biomedical research organizations. Their general mission is to conduct biomedical and behavioral research that will lead to the better health of the American people. For the NIH/ADAMHA investigator, this agency mission prescribes the exploration of ideas, the communication of ideas and information to colleagues, and a responsibility for the prompt and accurate publication of findings. Under the FTTA, 15 U.S.C. at § 3710a(a)(2), technology transfer, consistent with mission responsibilities, is also a responsibility of each laboratory science and engineering professional. To support their mission, NIH/ADAMHA have developed an interdisciplinary and synergistic research environment that promotes the free exchange of ideas and information. In order to safeguard the collegiality and integrity of, as well as public confidence in, the NIH/ADAMHA research programs, the following cooperative research and technology transfer policies have been adopted.

1. RESEARCH FREEDOM

NIH/ADAMHA investigators generally are free to choose the subject matter of their research, consistent with the mission of their Institute and the research programs of their Laboratories. No CRADA or license agreement may contravene this freedom.

2. RESEARCH POLICY

NIH/ADAMHA research results generally are disseminated freely through publication in the scientific literature and presentations at public fora. Brief delays in this dissemination of research results may be permitted under a CRADA as necessary in order to file corresponding patent or other intellectual property applications. NIH/ADAMHA consider the filing of such applications to be an important component of their research efforts.

* **Questions or comments about this Statement and requests for updated versions should be directed to the NIH Office of Technology Transfer at (301) 496-0750. This statement is effective on an interim basis, and will be revised after October 1, 1989.**

3. COOPERATIVE RESEARCH AND DEVELOPMENT UNDER A CRADA

As defined by the FTTA, 15 U.S.C. at § 3710a(d)(1), a CRADA means any agreement between one or more Federal laboratories and one or more non-Federal parties, under which the Government provides personnel, services, facilities, equipment or other resources (but not funds), and the non-Federal parties provide funds, personnel, services, facilities, equipment or other resources toward the conduct of specified research or development efforts. Cooperative research and development activities are intended to facilitate the transfer of federally funded research and development for use by State and local governments, universities and the private sector, particularly small businesses.

4. NIH/ADAMHA CRADAS

As adopted by NIH/ADAMHA, a CRADA is a standardized agreement intended to provide an appropriate legal framework for, and to expedite the approval of, cooperative research and development projects. The use of CRADAs is encouraged for cooperative efforts because they permit NIH/ADAMHA to accept, retain, and use funds, personnel, services, and property from collaborating parties and to provide personnel, services, and property to collaborating parties. NIH/ADAMHA may permit their investigators to enter into CRADAs with collaborators who will make a significant intellectual contribution to the research project undertaken or who will contribute essential research materials or technical resources not otherwise reasonably available. Although NIH/ADAMHA welcome contributions to their gift funds for research purposes, they do not view CRADAs as a general funding source or a mechanism for sponsored research. This approach to implementing the FTTA has been chosen in order to maintain the public's confidence in NIH/ADAMHA through maintaining an independence from reliance on industry funding.

5. SELECTION OF COLLABORATORS UNDER A CRADA

Collaborations under a CRADA may be suggested by potential collaborators or by NIH/ADAMHA investigators. Generally, the decision to initiate the approval process for a CRADA is made by the involved NIH/ADAMHA investigator and laboratory chief based on scientific considerations and the desire for the public to benefit from the commercialization of particular NIH/ADAMHA research. For some cooperative projects, where the development and commercialization potential is more immediate relative to the basic research aspects, NIH/ADAMHA may seek a collaborator(s) which has both scientific expertise and commercialization capabilities. In certain areas of research, e.g., where the Government has the intellectual lead or where both scientific and commercialization capabilities are deemed essential at the outset, NIH/ADAMHA may competitively seek a collaborator through Federal Register notification. The PHS has also developed policy guidelines for ensuring fairness of access to PHS laboratories such as NIH and ADAMHA in the process of initiating and developing CRADAs.

6. PROPRIETARY OR CONFIDENTIAL INFORMATION AND MATERIALS

NIH/ADAMHA recognize that an effective collaborative research program may require the disclosure of proprietary information to NIH/ADAMHA investigators. Although agreements to maintain confidentiality are permitted under a CRADA, collaborators should limit their disclosure of proprietary information to the amount necessary to carry out the research plan of the CRADA. The mutual exchange of confidential information, e.g., patient data, should be similarly limited. NIH/ADAMHA also recognize that cooperative research may require the exchange of proprietary research materials. Such material may be used only for the purposes specified in the research plan set forth in the CRADA. All parties to the CRADA will agree to keep CRADA research results confidential to the extent permitted by law until they are published in the scientific literature or presented at a public forum.

7. TREATMENT OF DATA AND RESEARCH PRODUCTS PRODUCED UNDER A CRADA

The NIH/ADAMHA investigator and the collaborator will agree to exchange all data and research products developed in the course of research under a CRADA whether developed solely by NIH/ADAMHA, jointly with the collaborator, or solely by the collaborator. In general, tangible research products developed under a CRADA will be shared equally by the parties to the CRADA. All parties to a CRADA will be free to utilize such data and research products for their own purposes. Data and research products developed solely by the collaborator may be designated as proprietary by the collaborator when they are wholly separable from the data and research products developed jointly with NIH/ADAMHA investigators. However, except as may be afforded through intellectual property rights that require public disclosure of the protected subject matter (e.g., patents), NIH/ADAMHA will not agree to exclude others from utilizing or commercializing the data or research products developed solely by NIH/ADAMHA investigators or jointly with the collaborator under a CRADA.

8. OWNERSHIP AND LICENSING OF NIH/ADAMHA INTELLECTUAL PROPERTY RIGHTS

Pursuant to the FTTA, 15 U.S.C. at § 3710a(b)(2), a Federal laboratory is authorized to own and license patent rights to inventions made in whole or part by its employees under a CRADA. The term "invention" is defined at § 3703 (9) to mean any invention or discovery which is or may be patentable or otherwise protected under Title 35 or any novel variety of plant which is or may be protectable under the Plant Variety Protection Act (PVPA), 7 U.S.C. § 2321 et seq. The patent law, 35 U.S.C. at § 207, authorizes the ownership and licensing of intramural inventions. Executive Order 12591 at § 1(b)(1)(B) further authorizes the transfer of Government intellectual property rights. Although the FTTA speaks broadly of the transfer of "technology," NIH/ADAMHA do not have statutory authority to license (or to agree to limit dissemination of) technology developed in whole or in part by their investigators under a CRADA unless a patent, PVPA certificate or other intellectual property application has been filed for that technology. NIH/ADAMHA will retain the Government ownership interest in, and license rights to, all intellectual property rights to inventions developed solely through intramural research or developed in whole or in part by their investigators under a CRADA.

9. GENERAL LICENSING POLICY

NIH/ADAMHA recognize that under the FTTA and the patent licensing law to which it refers, Congress and the President have chosen to utilize the patent system as the primary mechanism for transferring Government inventions to the private sector. The importance of patents to commercialization in the biomedical field is further reflected by the Drug Price Competition and Patent Term Restoration Act of 1984 (P.L. 98-417). A fundamental principle of the patent system is that the owners of a patent have a time-limited "right to exclude others from making, using, or selling the [patented] invention." The reason for such a period of exclusivity is to encourage industry to invest the resources necessary to bring an invention from the discovery stage through subsequent development, clinical trials, regulatory approvals, and ultimately into commercial production. NIH/ADAMHA accordingly are willing to grant exclusive commercialization licenses under their patent or other intellectual property rights in cases where substantial additional risks, time and costs must be undertaken by a licensee prior to commercialization. Under a CRADA, NIH/ADAMHA are also willing to agree to grant exclusive commercialization licenses in advance to collaborators. NIH/ADAMHA will attempt, however, to license their intramural inventions nonexclusively in cases where an invention reflects a relatively more advanced stage in its commercial development, e.g., when an NIH/ADAMHA investigator invents a patentable new therapeutic use for a known and FDA-approved compound.

Federal laboratories are authorized to negotiate license agreements for Government-owned patent rights in intramural inventions pursuant to 35 U.S.C. § 207. Although § 207 does not apply to intellectual property license agreements authorized by the FTTA for inventions made under a CRADA, NIH/ADAMHA have adopted the following approach of § 207 for all license agreements:

> Each Federal Agency [may] ... grant nonexclusive, exclusive, or partially exclusive licenses under federally owned patent applications, patents, or other forms of protection ... on such terms and conditions ... as determined appropriate in the public interest.

NIH/ADAMHA have determined it to be appropriate in the public interest to grant nonexclusive research licenses and either exclusive or nonexclusive commercialization licenses to DHHS-owned intellectual property rights according to the plan discussed below.

10. GOVERNMENT INTELLECTUAL PROPERTY RIGHTS

For inventions developed wholly by NIH/ADAMHA investigators or jointly with a collaborator under a CRADA, NIH/ADAMHA are required by the FTTA at 15 U.S.C. § 3710a(b)(2) to retain at least a nonexclusive, irrevocable, paid-up license to practice the invention or to have the invention practiced throughout the world by or on behalf of the U.S. Government. When granting exclusive or partially exclusive licenses to NIH/ADAMHA intramural inventions, 35 U.S.C. § 208, as implemented by 37 C.F.R. § 404.7(2)(i), requires the reservation of similar Government rights. NIH/ADAMHA will not assert an ownership right in inventions made solely by a collaborator under a

CRADA, but will require the grant of a research license, as described below, to the Government for inventions made wholly by a collaborator under a CRADA.

11. RESEARCH LICENSES

NIH/ADAMHA will reserve the right under any CRADA and intellectual property license to grant nonexclusive licenses to make and to use the invention for purposes of research involving the invention itself, and not for purposes of commercial manufacture or in lieu of purchase as a commercial product for use in other research. The purpose of the research license is to facilitate basic academic research. NIH/ADAMHA intend to consult with any involved commercialization licensee(s) before granting research licenses to commercial entities.

12. COMMERCIALIZATION LICENSES

NIH/ADAMHA are willing to consider requests for nonexclusive or exclusive commercialization licenses to intellectual property rights to inventions developed under a CRADA or in the course of intramural research, pursuant to applicable statutes and regulations. Under a CRADA, NIH/ADAMHA generally will grant a time-limited option to negotiate, in good faith, the terms of a license that fairly reflects the relative contributions of the parties, the risks incurred by the collaborator and the costs of subsequent research and development needed to bring the results of CRADA research to the marketplace. NIH/ADAMHA contemplate the drafting of a model invention license to serve as the starting point for license negotiations. It is contemplated further that such a model will reduce negotiations essentially to matters of execution fees, royalty rates and minimum annual royalties. Royalty rates will be based on product sales and the rates conventionally granted in the field identified in the CRADA's research plan for inventions with reasonably similar commercial potential. Royalty rates generally will not exceed a rate within the range of 5% to 8% for exclusive commercialization licenses. Contingent royalty schemes based on, e.g., patent issuance or nonissuance, and clauses treating the stacking of royalties or packaging of other inventions developed under the CRADA may be provided. Exclusive licensees will be expected to reimburse NIH/ADAMHA for intellectual property related expenses, and may be permitted to offset such reimbursement against future product royalties.

13. NONEXCLUSIVE COMMERCIALIZATION LICENSES

Unless a request for exclusive commercialization license is made under a CRADA or submitted for an intramural invention, NIH/ADAMHA will attempt to license their inventions nonexclusively. Such nonexclusive licenses generally will follow the guidelines of 37 C.F.R. Part 404.

14. EXCLUSIVE COMMERCIALIZATION LICENSES

All NIH/ADAMHA exclusive commercialization licenses will require the submission by a prospective licensee of an acceptable development and commercialization plan as described by 35 U.S.C. § 209(a) and subsequent, periodic reports on utilization of the invention as described by § 209(f)(1). All such plans and reports will be treated in confidence and as privileged from disclosure under the Freedom of Information Act. Modification provisions as described by § 209(f)(2)–(4) may apply. In appropriate cases, NIH/ADAMHA may also reserve the right to grant separate exclusive commercialization licenses in various fields of use. The remaining provisions of 35 U.S.C. §§ 200–212 will also apply to licenses to NIH/ADAMHA intramural inventions.

NIH/ADAMHA also consider the following provisions for exclusive commercialization licenses to be necessary and appropriate in the public interest:

> (i) the exclusive licensee must pledge its reasonable best efforts to commercialize a licensed invention and the development and commercialization plan mentioned above may serve as the measure of such efforts;
>
> (ii) NIH/ADAMHA shall have the right, after notice and opportunity to cure, to terminate or render nonexclusive any license granted: (1) if the licensee is not reasonably engaged in research, development, clinical trials, manufacturing, marketing, sublicensing, or other activities reasonably necessary to the expeditious commercial dissemination of the licensed invention; or (2) when the licensee cannot reasonably satisfy unmet health and safety needs;
>
> (iii) in order to maximize the commercialization of the licensed invention in other fields of use not utilized by the exclusive licensee through ongoing development, manufacturing or sublicensing, NIH/ADAMHA reserve the right to require the licensee to grant sublicenses to responsible applicants, on reasonable terms, in such other fields of use, unless the licensee can reasonably demonstrate that such a sublicense would be contrary to sound and reasonable business practice and the granting of the sublicense would not materially increase the availability to the public of the licensed invention; and
>
> (iv) exclusive licensees to DHHS inventions, whether developed under a CRADA or through intramural research, must agree to not unreasonably deny requests for sublicense or cross-license rights from future CRADA collaborators when the possibility of acquiring such derivative rights is necessary in order to permit a proposed cooperative research project with NIH/ADAMHA to go forward, and the exclusive licensee has been given a reasonable opportunity to join as a party to the proposed CRADA.

15. COMPLIANCE UNDER A CRADA WITH OTHER POLICIES

For research conducted pursuant to a CRADA, collaborators must agree to comply with PHS, NIH and ADAMHA policies and guidelines concerning, e.g., human subjects research, the use of research animals including nonwild chimpanzees, recombinant DNA and other policy statements as may be promulgated from time to time.

16. PRICING

DHHS has responsibility for funding basic biomedical research, for funding medical treatment through programs such as Medicare and Medicaid, for providing direct medical care and, more generally, for protecting the health and safety of the public. Because of these responsibilities, and the public investment in the research that contributes to a product licensed under a CRADA, DHHS has a concern that there be a reasonable relationship between the pricing of a licensed product, the public investment in that product, and the health and safety needs of the public. Accordingly, exclusive commercialization licenses granted for NIH/ADAMHA intellectual property rights may require that this relationship be supported by reasonable evidence.

17. WAIVERS

NIH/ADAMHA will consider requests to modify any of the foregoing policies in special cases where public health exigencies or commercial situations warrant such a modification. Modifications dealing with business terms such as royalties are not decided by the NIH/ADAMHA investigators and should be discussed with the appropriate NIH/ADAMHA technology management personnel.

18. SPECIAL CONSIDERATION AND PREFERENCE UNDER A CRADA

NIH/ADAMHA will give special consideration to entering into CRADAs with small business firms and consortia involving small business firms; and will give preference to business units located in the United States that agree to manufacture substantially in the United States products that embody inventions developed in the course of research under CRADAs.

APPENDIX B
RESEARCH PLAN

TITLE OF CRADA: _____.

NIH/ADAMHA PRINCIPAL INVESTIGATOR: _____
and his/her Laboratory: _____.

COLLABORATOR PRINCIPAL INVESTIGATOR: _____
_____.

TERM OF CRADA: ____ ( ) years.

CONFLICTS OF INTEREST INFORMATION: Describe any relevant past, present or contemplated relationships between the NIH/ADAMHA Principal Investigator and

his/her Laboratory and the Collaborator in sufficient detail to permit reviewers of this CRADA to determine whether any conflicts of interest exist: _____
_____
_____
_____.

The Research Plan that follows this page should be concise but of sufficient detail to permit reviewers of this CRADA to evaluate the scientific merit of the proposed collaboration. The RP should explain the scientific importance of the collaboration and the research goals of NIH/ADAMHA and the Collaborator. The respective contributions in terms of expertise and/or research materials of NIH/ADAMHA and Collaborator should be summarized. Initial and subsequent projects contemplated under the RP, and the time periods estimated for their completion, should be described and pertinent methodological considerations summarized. Pertinent literature references may be cited and additional relevant information included. Include additional pages to identify the Principal Investigators of all other Parties to this CRADA.

SUPPLEMENT TO APPENDIX B
GUIDELINES FOR DRAFTING THIS RESEARCH PLAN

In order to assist in the drafting of an appropriate Research Plan and to facilitate its processing and approval at NIH/ADAMHA, the following supplement has been adopted. Please use as many additional pages as necessary in order to respond fully, and number responses consistent with the numbering below. These guidelines and explanatory notes are an incorporated part of this CRADA.

1. GOAL OF THIS CRADA

Explanatory Note: Identify (three to four sentences) the research goal(s) of this CRADA, including the respective research goals of the NIH/ADAMHA and Collaborator Principal Investigators. Explain why this project is important scientifically.

2. DETAILED DESCRIPTION OF THE RESEARCH PLAN

Explanatory Note: The primary purpose of this Research Plan is to permit careful monitoring of CRADA research projects by scientific and division directors of our institutes, centers and divisions. An additional purpose for the Research Plan is established by the Federal Technology Transfer Act of 1986 (FTTA). Under the FTTA, the Parties' obligations to each other in such areas as confidentiality and patent rights extend only to *"specified research or development efforts."* This statutory limitation will create the boundaries for license rights to inventions made under the CRADA. Appropriate care should be taken in drafting this Research Plan carefully and completely. The field(s) of use to which Article 8 of this CRADA pertains will be limited to the specified research or development efforts in view of the foregoing research goals. The Collaborator further should bear in mind that, although insubstantial changes in this Research Plan may be made by mutual written consent of the Principal Investigators under Article 3.4, substantial changes will require formal amendment under Article 14.6 in order to maintain entitlement to invention rights. Absent *compelling* justification for a failure to make the original Research Plan complete, amendments will *not* be made retroactive.

Therefore, please provide a description (two to five pages) of the intended Research Plan in sufficient detail to permit reviewers of the CRADA to evaluate the scientific merit of the proposed collaboration. The Research Plan should be described in detail in terms of *specific research projects*—not in terms of a general research program or research goals. Contemplated initial and subsequent projects should be summarized along with estimated time periods for their completion. These projects may be described sequentially in distinct phases contingent upon the success of earlier phases. Important methodological considerations should be noted, and citations to pertinent literature reference may be helpful.

3. Respective Contributions of the Parties

Explanatory Note: Under Paragraph 4 of the NIH/ADAMHA "Policy Statement" (see Appendix A), CRADAs are authorized only with collaborators who will make a significant intellectual contribution to the research project undertaken, or who will contribute essential research materials or technical resources not otherwise reasonably available to NIH/ADAMHA. CRADAs are not viewed by NIH/ADAMHA as a general funding source or as a mechanism for sponsored research. Thus, unless essential materials or technical resources are involved, the Research Plan must indicate clearly that a true intellectual collaboration will take place. With regard to the detailed Research Plan described above, identify in detail by Party and by Principal Investigator the respective contributions of research, development, analysis, expertise, research materials, time, etc. to be committed to the various specified research projects and their component steps.

4. Abstract of the Research Plan for Public Release

Explanatory Note: In order to fulfill their obligations regarding NIH/ADAMHA activities to the public, to Congress and to the scientific community, NIH/ADAMHA intend to make an abstract of this Research Plan available upon request. To protect the legitimate concerns of the Collaborator as to its research agenda, the Collaborator is requested to assist in and carefully review this abstract. Signature of this CRADA by the Collaborator shall be deemed to be agreement by the Collaborator that NIH/ADAMHA may disclose this abstract publicly.

5. RELATED CRADAS

The Collaborator should identify by Title, Principal Investigator and Institute all other CRADAs that it has with NIH/ADAMHA. The NIH/ADAMHA Principal Investigator should similarly identify all CRADAs that his or her laboratory has with this or any other Collaborator.

6. RELATED MTAS

The NIH/ADAMHA Principal Investigator carefully must review his or her laboratory files and attach to the clearance form for this CRADA any material transfer agreements from any source that provided research materials used in earlier projects that relate directly or indirectly to this CRADA, *or* that provided research materials used to develop any materials to be studied or utilized in this CRADA. The ICD Technology Development Coordinator should similarly review any central material transfer agreement files and attach relevant agreements.

7. RELATED PATENT APPLICATIONS AND PATENTS

The NIH/ADAMHA Principal Investigator and Technology Development Coordinator should identify by title and serial number any ICD patent applications and patents that are directly or indirectly related to the subject matter of this CRADA.

8. AVOIDANCE OF CONFLICT OF INTERESTS AND ASSURANCE OF FAIR ACCESS

Explanatory Note: NIH/ADAMHA have implemented the FTTA with strict attention to Federal conflict of interest and ethic laws, as well as various Departmental and NIH/ADAMHA regulations. Additionally, the Public Health Service has issued guidelines for PHS agencies in order to assure fair access to our laboratories and consideration for CRADAs. Completion and signature certification of the following conflict of interest disclosure and fair access assurance form by the NIH/ADAMHA Principal Investigator only is mandatory prior to review of a proposed CRADA by the CRADA Subcommittee.

Financial and Staffing Contributions of the Parties

Exceptions or Modifications to This CRADA

INDEX

A

AdaNET software repository, 86
Advanced Interventional Systems, Inc. (AIS), 92, 109
Advanced Technology Program (ATP), 110–11
AEP (Applications Engineering Program), 57, 109
Aerospace and Defense Science Magazine, 30
Aerospace Research Applications Center. *See* Center for Aerospace Information (CASI)
AGRICOLA, 68
Agricultural Extension Service, 25, 26
Agricultural Research Service (ARS), 4
    conferences/trade shows organized by, 65
    contacts and laboratories for, 121–27
    cooperative research and development agreements and, 27, 98–99
    laboratories run by, 38
    licensing opportunities and, 26
    patent licensing and, 75, 76
    TEKTRAN, 27, 67
Agriculture, U.S. Department of. *See also* U.S. Forest Service
    contacts and laboratories for, 121
    creation and unification of, 13
    laboratories run by, 38
    land grant university system, 9, 13
    National Agricultural Library, 26, 27, 68
    Office of International Cooperation and Development, 71
    Online, 68
    policies and structure of, 26–27
    as a source of information, 66, 67, 68, 71, 106–7
Air Force, U.S., 17
    brokers, 60
    contacts and laboratories for, 130–31
    contractor-owned technology and, 90
    cooperative research and development agreements and, 30, 99
    Engineering and Service Laboratory, 4
    Office of Research and Technology Applications and, 30
    patent licensing and, 76, 81
    Regulation 80-27 (Domestic Technology Transfer), 17, 30
Air Products and Chemicals, Inc., 109
Alcohol, Drug, and Mental Health Administration (ADAMHA), 31, 65, 70
    sample documents, 159–82
Alternative Treatment Technology Information Center (ATTIC), 34, 69
American Superconductor Corp., 89
American Wind Energy Association (AWEA), 58
Ames Laboratory, 61, 70
Ames Research Center, 80
Apple Computer, 86
Applications Engineering Program (AEP), 57, 109
APS Materials, 80
Argonne National Laboratory (ANL), 50, 59, 61, 66, 89, 147
Army, U.S.
    AUTOCAD, 85
    CADVANCE, 85
    Chemical Research, Development, and Engineering Center, 76

183

Civil/Construction Engineering Research Laboratories (CERL), 5, 70–71, 85, 99, 147
consulting and, 109
contacts and laboratories for, 131–33
cooperative research and development agreements and, 29, 99
Laboratory Command (LABCOM), 29
Material Command, 29
Materials Technology Laboratory (MTL), 110
Office of Research and Technology Applications and, 29
patent licensing and, 76, 80
Signal Corps, 14
Small Business Innovative Research (SBIR) program, 29
ARS. See Agricultural Research Service
Ashton—Tate, 83
Atomic Energy Commission, 21, 101
ATP (Advanced Technology Program), 110–11
Aura Medical Systems, Inc., 88
Aura Systems, 87–88
AUTOCAD, 85
Automated Manufacturing Research Facility (AMRF), 106

B

Bankhead—Jones Act of 1935, 14
Bayh—Dole Act, 9–10
BEVA (Building Element Vector Analysis), 85
BLAST (Building Loads and System Thermodynamics), 85
Bob's Candies, 109
Boiler technology, 9
Bon Del Manufacturing Co., 11
Brokers, 42
 advisory boards as, 57–58
 Commerce Department, 58
 Defense Department, 60
 Energy Department, 60–61
 Federal Laboratory Consortium, 56–57, 59
 fees, 57
 function of, 55–56
 Industrial Applications Centers, 57
 inreach/technology pull, 55–56
 NASA, 56, 57, 59
 National Environmental Technology Applications Corporation, 60
 National Institute of Standards and Technology, 56, 106
 National Renewable Energy Laboratory, 57–58, 81, 106
 navy, 56
 Office of Research and Technology Applications and, 56
 outreach/technology push, 56
 success factors for, 61–62
Brookhaven National Laboratory, 101
Brown Foundation, 86
BRS, 68
Building Element Vector Analysis (BEVA), 85
Building Loads and System Thermodynamics (BLAST), 85
Bureau of Mines, 22
 contacts and laboratories for, 138
 function of, 39
 policies and structure of, 33
Bureau of Reclamation
 contacts and laboratories for, 138–39
 cooperative research and development agreements and, 32, 99
 function of, 39
 Office of Research and Technology Applications and, 32
 patent licensing and, 81
Byrd, Robert, 67

C

CADVANCE, 85
Carey, William D., 44
CD-ROM, 68

Center for Aerospace Information (CASI), 66, 69, 72
Centers for Disease Control (CDC), 31
Centers for the Commercial Development of Space (CCDS), 59
Central Institute for the Deaf, 4
Ceramic anode (CERANODE), 80
CERL. *See* Civil/Construction Engineering Research Laboratory
Cermet, 89
Chapman Research Group, Inc., 11
ChemMap, 85
Civil/Construction Engineering Research Laboratories (CERL), 5, 70–71, 85, 99, 147
Clear Creek Independent School District (Texas), 86
Clearinghouse Electronic Mail and Resource Directory, 66–67
Clearinghouse for State and Local Initiatives on Productivity, Technology, and Innovation, 28, 58
Commerce, U.S. Department of
 brokers, 58
 Clearinghouse for State and Local Initiatives on Productivity, Technology, and Innovation, 28, 58
 conferences/trade shows organized by, 64, 65
 contacts and laboratories for, 127–30
 cooperative research and development agreements and, 28
 Federal Technology Transfer Series by, 64, 65
 laboratories run by, 38
 National Institute of Standards and Technology (NIST), 8, 28, 38–39, 66, 68, 69, 106
 National Oceanographic and Atmospheric Agency (NOAA), 28, 29, 38
 National Technical Information Service (NTIS), 8, 28, 71
 National Telecommunications and Information Administration (NTIA), 28
 policies and structure of, 27–29
Commerce Business Daily, 44, 69, 70, 100
Computer Aided Logistics (CALS), 71
Computerized Visual Communication (C-VIC), 92, 93
Computer Software Management and Information Center (COSMIC), 57, 85–86, 144
Conflict-of-interest guidelines, 107–8
Connecticut Department of Economic Development, 65
Connecticut Innovations, 65
Construction/Civil Engineering Research Laboratory (CERL), 5, 70–71, 85, 99, 147
Consulting/consultants
 number of federal researchers involved in, 108–9
 use of, 107
Contractor-owned technology
 Air Force and, 90
 examples of transfer of, 87–88
 small companies and, 90
 technology transfer from GOCOs, 88–89
Cooperative research and development agreements (CRADAs)
 access problems to, 47
 Agricultural Research Service and, 27, 98–99
 Air Force and, 30, 99
 Army and, 29, 99
 Bureau of Reclamation and, 32, 99
 Commerce Department and, 28
 creation of, 15
 definition of, 97–98
 Energy Department and, 99
 examples of, 98–99
 Federal Highway Administration and, 34
 Food and Drug Administration and, 32
 growth of, 16, 17, 20
 how to start the process and tips for using, 99–100
 NASA and, 35
 National Institutes of Health and, 17, 31–32, 98, 159–82
 Navy and, 30
 new funding sources for agency R & D programs and, 22
 patent licensing and, 80, 81
 sample documents, 159–82

U.S. Forest Service and, 27
Veterans Affairs and, 34
Cooperative State Research Service (CSRS), 26
Copyrights
　GOCO software copyright policy, 84
　problems over software, 45, 83–84
Corning Glass Works, 72
COSMIC (Computer Software Management and Information Center), 57, 85–86, 144
CRADAs. *See* Cooperative research and development agreements
CRISP Intramural Research Index, 70
Cuyahoga Community College, 70
C-VIC (Computerized Visual Communication), 92, 93

D

Data bases, information through, 66–69
DayChem Corp., 81
dBase, 83
DD Form 2345 Militarily Critical Technical Data Agreement, 68–69, 148
Defense, U.S. Department of (DOD), 10
　brokers, 60
　conferences/trade shows organized by, 64, 65
　contacts and laboratories for, 130–35
　Federal Laboratory Consortium (FLC) of, 14, 16
　laboratories run by, 37–38, 102
　patent licensing and, 76
　policies and structure of, 29
　Regulation 3200.12-R-4, 29
　as a source of information, 66
　Technology Transfer Laboratory Consortium of, 14
Defense Logistics Agency (DLA), 69
Department of. *See under name of*, e.g., Energy, U.S. Department of
Desktop Vocational Assistant Robot (DeVAR), 92, 93

DIALOG, 66, 68
Diamonex, Inc., 109
DOD. *See* Defense, U.S. Department of
DOE. *See* Energy, U.S. Department of
Domestic Technology TransFair, 64
Domestic Technology Transfer (Air Force regulation), 17, 30
Domestic Technology Transfer Fact Sheet (Navy) 30, 70
DOT. *See* Transportation, U.S. Department of
Dravo Lime Co., 61
Du Pont, 110

E

Edge Technologies, Inc., 61
Electronic Bulletin Board, Public Health Service, 68
Electronic Courseware Systems (ECS), Inc., 85
Electronics and Defense/Quest (TRW), 43
E-mail systems, 66–69
Embrex, Inc., 98–99
Energy, U.S. Department of (DOE)
　BLAST (Building Loads and System Thermodynamics), 85
　brokers, 60–61
　ChemMap, 85
　conferences/trade shows organized by, 64
　consulting and, 108
　contacts and laboratories for, 135–37
　conversion of wood waste and, 5
　cooperative research and development agreements and, 99
　Database (EDB), 68
　laboratories run by, 37, 38, 101
　National Renewable Energy Laboratory (NREL), 3, 57–58, 70, 81, 85
　patent licensing and, 76
　policies and structure of, 31
　problems with selling new ideas, 44
　as a source of information, 70
　technology transfer and, 88–89

Engineering News Record, 34
Environmental Protection Agency (EPA)
    Alternative Treatment Technology Information Center (ATTIC), 34, 69
    contacts and laboratories for, 140–42
    laboratories run by, 38, 39
    National Environmental Technology Applications Corporation (NETAC), 34
    policies and structure of, 34
    toxic waste remediation techniques and, 5
Executive Order 12591, 10, 16
Exxon, 50

F

"Facilitating Access to Science and Technology," 10
FDA. *See* Food and Drug Administration
Federal Applied Technology Data Base, 66
Federal Aviation Agency, 39
Federal Computer Products Center (FCPC), 84, 144
Federal Emergency Management Agency, 142
Federal Highway Administration (FHWA), 33–34
Federal Laboratory Consortium (FLC)
    as brokers, 56–57, 59
    Clearinghouse Electronic Mail and Resource Directory, 66–67
    conferences/trade shows organized by, 64, 65
    contacts for, 146–47
    function of, 14, 16
    Office of Research and Technology Applications and, 40
    as a source of information, 66–67
Federal Register, 44, 69, 77, 78
Fish and Wildlife Service
    Office of Research and Technology Applications and, 32–33
    research field stations of, 39

FLC. *See* Federal Laboratory Consortium
Food and Drug Administration (FDA)
    cooperative research and development agreements and, 32
    laboratories run by, 39
    Office of Research and Technology Applications and, 32
    shatter-resistant lenses and, 80
Forest Products Laboratory, 27, 102
Forest Service Information Network (FS INFO), 67
Foster Grant Corp., 80
Franklin Institute, 9
Freedom of Information Act, 46, 78, 100

G

General Accounting Office (GAO), 20, 46, 76
General Dynamics Corp., 107
Gerola, Humberto, 92–93
Goddard Space Flight Center, 21, 59
Government. *See also* Incentives, government
    technology transfer and lessons learned by, 49–50
Government Microcircuits Applications Conference (GOMAC), 64, 147
Government-owned, contractor-operated (GOCO) national laboratories
    cooperative research and development agreements and, 16
    function of, 10
    software copyright policy, 84
    technology transfer from, 88–89

H

Harry Diamond Laboratories (HDL), 80, 109
Hatch Act of 1887, 13–14

HAZARD 1, 84
Health and Human Services, U.S. Department of
  conferences/trade shows organized by, 64
  contacts and laboratories for, 137-38
  laboratories run by, 38, 39
  policies and structure of, 31-32
High-Tech Update, 70
Horvath, Carlos, 63
House Committee on Science, Space, and Technology, 17
Human Genome Project, 99
Hygienic Laboratory, 14

I

Incentives, government
  access to restricted technology, 22
  new funding sources for agency R & D programs, 22
  royalty sharing, 22-23
  support of agency R & D programs, 21-22
Incentives, industry
  access to federally developed technology, 19-20
  cooperative research ventures, 20
  new licensing opportunities, 21
  reduction of R & D costs, 20
  use of facilities, researchers, and store of knowledge, 21
Industrial Applications Centers (IACs), 10, 42, 57
Industrial Guest Investigators (IGIs), 105
Industrial Research Associate Program, 106
Industrial Research Institute (IRI), 64, 65
Industry. *See also* Incentives, industry
  technology transfer and lessons learned by, 49-50
Information, sources of
  data bases and E-mail systems, 66-69
  direct response services, 70-71
  how to obtain, 63, 64
  federal laboratory participation in trade shows, 64-66

international, 71
print and other media as, 69-70
private sector, 71
Information centers, list of, 144-45
Inreach/technology pull, 55-56
Intelligent Physics Tutor, 86
Interior, U.S. Department of
  contacts and laboratories for, 138-39
  laboratories run by, 38, 39, 102
  policies and structure of, 32-33
Inventions. *See also* Patent licensing
  commercialization of, 15
Iowa State University, 61, 70
ISICAD, Inc., 85, 99
ITT Gilfillan, 10

J

Jet Propulsion Laboratory, 83, 91-92, 107
Johns Hopkins University, Applied Physics Laboratory of, 21
Johnson Space Center, 67, 86
Justice, U.S. Department of, 138

K

KangaROOS USA, Inc., 72
Kees—Goebel Medical Specialists, 64
Kelman, Charles, 87-88
KELMAST, 87-88
Kinetic Controls, 91

L

Lablink, 67
Laboratories
  access problems, 48

agency policies and means of access, 102–3
description of federal, 37–39, 101–2
differences between, 40–41
how to obtain entry into, 41–42
implementation problems for, 46–47
participation in trade shows, 64–66
Land grant university system, 9, 13
Langley Research Center, 59, 63, 91, 107
Laudenslager, James, 92
Lawrence Berkeley Laboratory, 85
Lawrence Livermore National Laboratory (LLNL), 89, 110
Lee Iaccoca Institute, 84
Lehigh University, 84
Lewis and Clark expedition, 9
Lewis Research Center, 70, 109
Licensing. *See also* Patent licensing
  problems with, 47
Licensing opportunities
  Agricultural Research Service and, 26
  creation of new, 21
  NASA and, 35
  U.S. Forest Service and, 26
LifeNet, 67
Life Technologies, Inc. (LTI), 99
Lima Technical College, 60
Lingraphica, 92
Logical Technical Services (LTS) Corp., 80
Los Alamos National Laboratory, 61

M

March-in clauses, 79
Marine Hospital Service (NY), 14
Marshall Space Flight Center, 59
Martin Marietta Energy Systems, Inc., 88–89
Materials Technology Laboratory (MTL), 110
Medical Free Electron Laser Centers, 30
Michigan Technological University (MTU), 59
Midwest Research Institute (MRI), 57, 89
Midwest Research Institute Ventures (MRIV), 57–58, 81, 89, 147

Minnesota Mining & Manufacturing Co., 4
Mississippi Technology Transfer, 59
Morgan State University, 59
Morrill Act of 1862, 13

N

NASA. *See* National Aeronautics and Space Administration
NASTRAN, 5, 65, 86
National Advisory Commission on Aeronautics, 14, 21
National Aeronautics and Space Act of 1986, 34
National Aeronautics and Space Administration (NASA)
  Applications Engineering Program (AEP), 57, 109
  brokers, 56, 57, 59, 107
  Center for Aerospace Information (CASI), 66, 69, 72
  Centers for the Commercial Development of Space (CCDS), 59
  coatings for helmet visors and, 4
  conferences/trade shows organized by, 64, 65
  contacts and laboratories for, 143
  cooperative research and development agreements and, 35
  digital hearing aid and, 4
  digital image processing and, 4
  Industrial Applications Centers (IACs), 42, 57
  Intelligent Physics Tutor, 86
  laboratories run by, 37, 38, 102
  licensing opportunities and, 35
  LifeNet, 67
  mine safety applications and, 22
  NASTRAN, 5, 65, 86
  Office of Commercial Programs, 59
  Patent Abstracts Bibliography (*NASA PAB*), 69, 77
  patent licensing and, 75, 76

policies and structure of, 34–35
royalty sharing and, 35
as a source of information, 66, 69, 71, 72, 107
SPACE TECH Lens, 80
spinoff applications from, 11, 34–35, 91–92
Tech Briefs, 8, 35, 69, 71, 72, 77, 86
Technology Utilization Network System (TUNS), 56, 67
Technology Utilization Offices (TUOs), 56
Technology Utilization Program of, 13, 14, 25

National Agricultural Library (NAL), 26, 27, 68
National Bureau of Standards. *See* National Institute of Standards and Technology
National Cancer Institute, 76
National Center for Atmospheric Research, 39
National Clonal Germ Plasm Repositories, 102
National Competitiveness Technology Transfer Act of 1989 (P.L. 101–189), 10, 16, 31, 116
National Environmental Technology Applications Corp. (NETAC), 34, 60, 147
National Fire Protection Association, 84
National Heart, Lung, and Blood Institute, 109
National Highway Institute, 33
National Institute of Mental Health (NIMH), 64
National Institute of Standards and Technology (NIST), 28
  Advanced Technology Program (ATP), 110–11
  Automated Manufacturing Research Facility (AMRF), 106
  brokers, 56, 106
  CARB Biological Macromolecule Crystallization, 68
  function of, 38–39
  HAZARD 1, 84
  Industrial Research Associate Program, 106
  manufacturing technology centers of, 48

Reference Data Program, 68
  Special Publication 763, 69
  Standard Reference Data Program, 85
  Supertrapp, 85
  Technical Information Databases, 68
National Institutes of Health (NIH), 4
  conferences/trade shows organized by, 65
  contacts and laboratories for, 137–38
  cooperative research and development agreements and, 17, 31–32, 98, 159–82
  creation of, 14
  fairness issues, 47
  laboratories run by, 39
  Office of Technology Transfer (OTT), 31
  patent licensing and, 75
  Patent Policy Board, 31
  royalty rates, 78
  sample documents, 159–82
  as a source of information, 70
National Oceanographic and Atmospheric Administration (NOAA), 28
  contacts and laboratories for, 128–30
  function of, 29, 38
  Tech Briefs, 29
National Park Service, 39
National Plant Germ Plasm System, 102
National Renewable Energy Laboratory (NREL)
  as brokers, 57–58, 81, 106
  consulting and, 108
  new products of, 3
  software, 85
  spinoff companies, 92
  Technical Assistance Service (TAS), 70
  technology available for licensing, 89
National Science Foundation (NSF), 38, 39, 65, 143
National Seed Storage Laboratory, 102
National Technical Information Service (NTIS), 8, 28
  address for, 145
  conferences/trade shows organized by, 65
  Federal Computer Products Center (FCPC), 84
  Soil Conservation Service (SCS) Computer Software, 85
  as a source of information, 65, 66, 71

National Technology Transfer Center (NTTC), 67, 147
National Telecommunications and Information Administration (NTIA), 28, 130
Naval Underwater Systems Center, 56
Navy, U.S.
  brokers, 56
  conferences/trade shows organized by, 64
  contacts and laboratories for, 133–35
  cooperative research and development agreements and, 30
  Domestic Technology TransFair, 64
  Domestic Technology Transfer Fact Sheet, 30, 70
  Office of Naval Technology, 30
  Office of Research and Technology Applications and, 30
  patent licensing and, 76
  technology ferret program, 56
NERAC, 42
New York State Science and Technology Foundation, 99
NIH. *See* National Institutes of Health
NIST. *See* National Institute of Standards and Technology
NOAA. *See* National Oceanographic and Atmospheric Agency
Northern Regional Research Center (NRRC), 106–7
NREL. *See* National Renewable Energy Laboratory
NTIS. *See* National Technical Information Service

O

Oak Ridge National Laboratory, 59, 61, 88, 101
Office of Commercial Programs, 59
Office of Information (OI), 68
Office of International Cooperation and Development, 71
Office of Invention Development, 69
Office of Naval Technology, 30
Office of Research (Veterans Affairs), 34
Office of Research and Technology Applications (ORTA)
  Air Force and, 30
  Army and, 29
  brokers, 56
  Bureau of Reclamation and, 32
  creation of, 15, 40
  Fish and Wildlife Service and, 32–33
  Food and Drug Administration and, 32
  function of, 40–41, 56
  growth of, 20
  National Oceanographic and Atmospheric Administration and, 29
  Navy and, 30
  U.S. Forest Service and, 27
  U.S. Geological Survey and, 33
Office of Research Policy and Technology Transfer, 33
Office of Technology Applications, 33
Office of Technology Transfer (OTT), 31
Ohio Technology Transfer Organization (OTTO), 60
Oklahoma Vocational-Technical Department, 59
Omnibus Trade and Competitiveness Act (1989), 84
ORTA. *See* Office of Research and Technology Applications
Outreach/technology push, 56

P

Pacific Northwest Laboratory, 89
Passive Solar Industries Council (PSIC), 58
Patent Abstracts Bibliography (*NASA PAB*), 69, 77
Patent and Trademark Amendments of 1980 (P.L. 96–517), 9–10
Patent and Trademark Office, 75
Patent licensing
  examples of successful, 79–81
  exclusive versus nonexclusive, 77
  march-in clauses, 79

non-civil servant inventions resulting from federal funding, 79
problems over, 45, 47
process, 76–78
royalties and, 75
royalty rates, 78
sources for, 81
statistics on, 75–76
Pennzoil Products Co., 86
Perceptive Systems, Inc., 92
Performance Technologies, Inc., 88
Perma Charge Corp., 89
Pharmaceutical Manufacturers Association, 31
Programmable implantable medicinal supply (PIMS), 21, 43
Project Share, 33
Public Health Service (PHS)
  Electronic Bulletin Board and, 68
  patent licensing and, 76
P.L. 96–480. See Technology Innovation Act of 1980/Stevenson—Wydler Act
P.L. 96–517. See Patent and Trademark Amendments of 1980
P.L. 99–502. See Technology Transfer Act of 1986
P.L. 101–189. See National Competitiveness Technology Transfer Act of 1989

R

R and D Magazine, 70
Reagan, Ronald, 10, 16
Regional Technology Transfer Centers (RTTCs), 57
  addresses for, 145
  as brokers, 42
  function of, 10
  as a source of information, 66
Renewable Energy Institute (REI), 58
Research and development (R & D), role of federal, 7–11
Research Results Data Base (RRDB), 68
Research Triangle Institute (RTI), 57

Rome Air Development Center, 99
Royalty rates, 78
Royalty sharing, 22–23
  NASA and, 35
RTTCs. *See* Regional Technology Transfer Centers
Rural Enterprises, 57

S

Sandia National Laboratories, 61, 89
Savannah River Co., 61
Science Citation Index, 71
Science Park Development Corp., 65
Scientific and Technical Information Facility (STIF), 66, 69
SDIO. *See* Strategic Defense Initiative Organization
Silver Platter, 68
Small Business Administration, 57
Small Business Development Centers (SBDCs), 59, 66
Small Business Innovation Research (SBIR), 29, 40, 90, 110, 146
Smith—Lever Act, 14
Smithsonian Institution, 143–44
Software
  copyright problems, 45, 83
  currently available and sources, 84–86
  GOCO software copyright policy, 84
  laws and regulations, 83
  Training Technology Act and, 84
Soil Conservation Service (SCS) Computer Software, 85
Solar Energy Industries Association (SEIA), 58
Solar Rating and Certification Corporation (SRCC), 58
Spinback, technological, 22
Spinoff companies
  example of successful, 92–93
  how to facilitate the process, 93
  NASA, 11, 34–35, 91–92
  NREL, 70

SRI International, 9
Steele, Richard, 93
Stennis Space Center, 59
Stevenson-Wydler Technology Innovation Act of 1980 (Technology Innovation Act of 1980/P.L. 96–480), 9, 15, 40, 115
Strategic Defense Initiative Organization (SDIO)
   as brokers, 58
   contact for, 147
   as a source of information, 66
   Technology Applications Information System (TAIS), 30
Supertrapp, 85

T

TAIS. *See* Technology Applications Information System
Tech Briefs (NASA), 8, 35, 69, 71, 72, 77, 86
Tech Briefs (NOAA), 29
Technical Assistance Service (TAS), 70
Technical Volunteer Service (TVS), 109
Technology
   pull, 55–56
   push, 56
Technology Applications Information System (TAIS)
   function of, 68–69
   as a part of SDIO, 30, 66
   as a source of information, 66
Technology for U.S. Industry, 70
Technology Innovation Act of 1980 (P.L. 96–480/Stevenson-Wydler Act), 9, 15, 40, 115
"Technology Opportunities" (U.S. Forest Service), 27
Technology Support Packages (TSPs), 69
Technology Targeting Database, 57, 66
Technology transfer
   benefits of, 10–12
   brokers, 42, 55–62
   checklist, 151–55
   contacts, 121–48

   examples of cost savings, 4–5
   examples of new and improved products, 3–4
   future for, 116–18
   historical perspective of, 13–17
   myths, 43–44
   problems encountered with, 44–48
   role of federal research and development, 7–11
   sample documents, 159–82
Technology Transfer Act of 1986 (P.L. 99–502)
   purpose of, 3, 10, 14, 15
   review of, 115–16
   royalties and, 22
"Technology Transfer Agreements between Industry and ARS: A Plain Language Guide," 27
Technology Transfer Conferences, Inc., 58, 64
Technology Transfer Information Center, 27
Technology Transfer Laboratory Consortium, 14
"Technology 2000," 35, 65
Technology Utilization Network System (TUNS), 56, 67
Technology Utilization Offices (TUOs), 56
Technology Utilization Program, 13, 14, 25
TEKTRAN, 27, 67
Tennessee Center for Research and Development, 61
Tennessee Valley Authority, 144
Texas Advanced Technology Program, 86
Tolfa Corp., 92
Training Technology Act, 84
Transportation, U.S. Department of (DOT)
   contacts and laboratories for, 139–40
   Federal Highway Administration (FHWA), 33–34
   laboratories run by, 38, 39
   Office of Research Policy and Technology Transfer, 33
   Urban Mass Transit Administration (UMTA), 33
Transportation Research Center, 39
TRW, 43

U

Unified Technologies Center (UTC), 70
Unisys, 64
U.S. Composites, Inc., 110
U.S. Conference of Mayors, 57
U.S. Department of Agriculture (USDA). *See* Agriculture, U.S. Department of
U.S. Forest Service
  budget and opportunities offered by, 26
  contacts and laboratories for, 127
  cooperative research and development agreements and, 27
  Forest Products Laboratory, 27, 102
  Forest Service Information Network (FS INFO), 67
  laboratories run by, 38, 102
U.S. Geological Survey (USGS)
  contacts and laboratories for, 139
  creation of, 14
  Office of Research and Technology Applications and, 33
University of California, Berkeley, 89
University of Chicago Development Corporation, 61
University of Georgia, 57, 85
University of Houston, 86
University of Illinois, 85
University of Pittsburgh, 34, 60
University of South Carolina, 61
University of Utah, 57
Urban Mass Transit Administration (UMTA), 33

V

Veterans Affairs (VA), U.S. Department of, 4
  contacts and laboratories for, 140
  cooperative research and development agreements and, 34
  laboratories run by, 38, 39
  Office of Research, 34
  software, 85
  as a source of information, 70
  spinoff companies and, 92–93
Videotapes, use of, 70

W

Ward, Ray, 10
Washington University (St. Louis), 4
Water Filter Company of America, 11
Water filtering system, development of, 10–11
Waters, William, 70
Weld Quality Monitor (WQM), 80
Westinghouse, 61
Wheeling Jesuit College, 67
Woodward Governor Co., 99
Wright—Patterson AFB, 60

Y

Yale University, 65